THE GREAT ARC

Into India
When Men and Mountains Meet
The Gilgit Game
Eccentric Travellers
Explorers Extraordinary
Highland Drove
The Royal Geographical Society's
History of World Exploration (*general editor*)
India Discovered
The Honourable Company:
A History of the English East India Company
Collins Encyclopaedia of Scotland
(*co-editor with Julia Keay*)
Indonesia: From Sabang to Merauke
Last Post: The End of Empire in the Far East
India: A History

THE
GREAT ARC

The Dramatic Tale
of How India Was Mapped and Everest Was Named

JOHN KEAY

HarperCollins*Publishers*

HarperCollins books may be purchased for educational, business, or sales promotional use. For information please write: Special Markets Department, HarperCollins Publishers Inc., 10 East 53rd Street, New York, NY 10022.

FIRST AMERICAN EDITION

Printed on acid-free paper

Library of Congress Cataloging-in-Publication Data

Keay, John.
    The great arc : the dramatic tale of how India was mapped and Everest was named / John Keay.
        p.cm.
    Includes index.
    ISBN 0-06-019518-5
    1. India. Great Trigonometrical Survey—History. 2. Geodesy—India—History—19th century. 3. Geodesy—Himalaya Mountains—History—19th century. 4. India—Surveys—History—19th century. 5. Himalaya Mountains—Surveys—History—19th century. I. Title.
    QB296.15 K43 2000
    526.1'0954—dc21

                                                        00-038292

00 01 02 03 04 ❖/RRD 10 9 8 7 6 5 4 3 2 1

For Julia

Contents

Illustrations

Index chart of the Great Trigonometrical Survey and Arc.
From C. R. Markham, *Memoir of the India Survey* (*c*.1870).

Everest's ill-fated operations in the Kristna-Godavari jungles in
1819–20, while Lambton was pushing north with the Great
Arc. From R. H. Phillimore, *Historical Records of the Survey of
India*, Volume IV (1954).

Lambton's grave at Hinganghat in Maharashtra. Photograph ©
Julia Keay.

Pre-Lambton theodolite. Reproduced courtesy of the British
Library.

Early surveying equipment. Reproduced courtesy of the British
Library.

Lambton's Great Theodolite. Photograph © Kirsty
Chakravarty. Courtesy of the Survey of India.

Levelling instruments. From Smyth and Thuillier, *Manual of
Surveying for India* (1851).

A perambulator equipped with a mileometer for measuring
distances. From Phillimore, *Historical Records of the Survey of
India*, Volume IV.

The eleventh-century Great Temple at Tanjore in Tamil
Nadu. Reproduced courtesy of the Survey of India.

William Lambton, painted by William Havell in 1822. From
Phillimore, *Historical Records of the Survey of India*, Volume
III (1954).

George Everest in 1843. From Phillimore, *Historical Records of
the Survey of India*, Volume IV.

Everest in later life. Reproduced courtesy of the Royal
Geographical Society.

Compensation bars. From George Everest, *An Account of a*

ix

Measurement of Two Sections of the Meridional Arc, Volume II (1847). Reproduced courtesy of the British Library.

The 1831 measurement of a base-line at Calcutta. From Phillimore, *Historical Records of the Survey of India*, Volume IV.

Hathipaon House. Photograph © Kirsty Chakravarty. Courtesy of the Survey of India.

Instrument designed by Everest for astronomical observations. Photograph © Kirsty Chakravarty. Courtesy of the Survey of India.

The survey pole marking the Himalayan terminus of the Great Arc. Reproduced courtesy of the British Library.

The rope-ladder at Srinagar in Garhwal. Oil painting by Thomas Daniell. Reproduced courtesy of the British Library.

Nanda Devi. Photograph by Hugh Routledge. Reproduced courtesy of the Royal Geographical Society.

Forty-foot towers of scaffolding were rigged as observation posts. From Everest, *An Account of a Measurement of Two Sections of the Meridional Arc*, Volume II. Reproduced courtesy of the British Library.

For the final triangulation across the plains, Everest designed sixty-foot towers built of masonry. From Everest, *An Account of a Measurement of Two Sections of the Meridional Arc*, Volume II. Reproduced courtesy of the British Library.

One of the Great Arc's towers survives at Begarazpur to the north of Delhi. From Phillimore, *Historical Records of the Survey of India*, Volume IV.

The Great Theodolite was hoisted to the top of the survey towers by a specially designed crane. From Everest, *An Account of a Measurement of Two Sections of the Meridional Arc*, Volume II. Reproduced courtesy of the British Library.

The 'Musoorie Hills'. Engraving by J. B. Allen (1845). Reproduced courtesy of Mary Evans Picture Library.

The 'Observatory' on The Ridge at Delhi. Reproduced courtesy of the British Library.

Plan of the observatories from which astronomical observations were conducted. From Everest, *An Account of a Measurement of Two Sections of the Meridional Arc*, Volume II. Reproduced courtesy of the British Library.

'Strange's Zenith Sector No. 1'. Photograph © Kirsty Chakravarty. Courtesy of the Survey of India.

Kangchenjunga. Photograph by W. H. Lonnell. Reproduced courtesy of the Royal Geographical Society.

The tribunal which settled the heights of the Himalayas: T. G. Montgomerie, A. S. Waugh, J. T. Walker and H. E. L. Thuillier. Reproduced courtesy of the British Library.

An artist's impression of Mount Everest leaves no doubt as to its pre-eminence. Reproduced courtesy of Mary Evans Picture Library.

For the photographer, variable weather and a cluster of rival peaks cloud the issue. Reproduced courtesy of the Royal Geographical Society.

Maps

A Note on Spellings

Some proper names in the text, especially place-names, will appear to be mis-spelled. This may be because a spelling acceptable to everyone has been hard to establish; or it may be because I have adopted some nineteenth-century spellings in order to be consistent with those used in the quoted extracts (e.g. 'Kistna' for 'Krishna', 'Siwaliks' for 'Shivaliks', 'Ganges' for 'Ganga', etc.). I trust that purists will show indulgence and that all the names are at least recognisable. In the case of the Bengali genius 'Radhanath Sickdhar', the spelling is that which he himself used, however improbable it now looks.

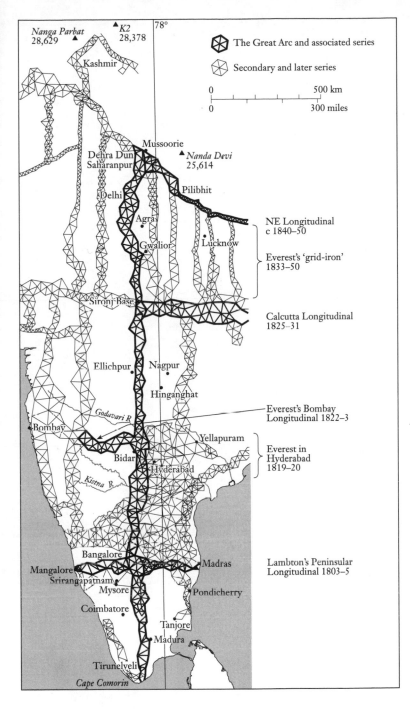

THE GREAT ARC AND ASSOCIATED SERIES

Foreword

Pressed about why Mount Everest is so named I would once, perhaps like most people, have come up with the explanation that it was as good a name as any. For a geographical feature of such obvious permanence and precedence, 'Everest' seems to say it all. Up there, more aloof from the bustle of life than anywhere else on earth, the raised snows proffer a pledge of peace, a promise of lasting repose, of being 'ever-at-rest'. Presumably some international body had ordained the name; or perhaps it was a translation of the mountain's local title. It scarcely mattered. Either way, it was perfectly acceptable.

But, as I now know, these suppositions were totally wrong. That the world's highest point is in fact called after George Everest, a controversial British Colonel who had never even seen the mountain, let alone climbed it, first dawned on me when I was writing a book about the exploration of Kashmir. Everest did not feature in the region, either as man or mountain, but an institution, dear to the Colonel's heart and known as the Survey of India, did. Most of Kashmir, including the Karakoram mountains, had first been measured and mapped by men of the Indian Survey. And the Survey being a government department within British India's bureaucratic Leviathan, it had generated copious records. To these I turned.

Descriptions of nineteenth-century map-makers hauling their instruments up peaks of unknown altitude proved excellent value. Pelted by hailstones, their tents ablaze from the lightning and their trail obliterated by blizzards, the men of the Survey would dig in and wait. Survival depended on merino drawers, Harris tweeds and alpaca overcoats. Their boots were leather, and they ate mostly rice. They might be marooned for weeks. Then, without warning, in the chill first light of a day when the cloud had unaccountably overslept in the valleys, their patience would at last be rewarded. Sailing a sea of cumulus beneath an azure sky, a line of glistening summits would loom remotely from the ether.

With luck, from two or more of these summits tell-tale pinpricks of light would advertise the presence of other survey teams. If the theodolite was up and ready, sightings would be taken, bearings recorded and signals exchanged. The job was done. A speedy retreat down to the fleshpots of basecamp followed. Then it was on and up to the next peak.

Or so it seemed; but taking such bearings enabled the surveyors to plot the positions and heights of the distant peaks only if the location of their own peak was already known. Otherwise it was like seeking directions from a street map without first having identified your whereabouts. Fixing the positions of the other peaks depended on knowing that of the one from which one observed; its global location in terms of the world's grid of longitude and latitude had to have been established, and so did its height above sea-level.

Obviously this information could have been obtained from prior observations taken at other vantage points in the rear. And those vantage points could in turn have been similarly established from somewhere in the foothills. But the Himalayas were hundreds of miles from any observatory capable of supplying a fix from astronomy; and they were even further from the sea, in terms of whose mean-level heights were expressed. Working back, then, somewhere there had to have been a starting point, a benchmark series of locations whose co-ordinates and heights had been deduced with a superior precision and were known with unimpeachable certainty.

I asked around, I dug out books, and the trail led back to the Great Arc – or, to give it its full title, 'the Great Indian Arc of the Meridian'. I had never heard of it; but the Arc was indeed that benchmark series of locations. It was like the trunk of a tree, the spinal column of a skeleton. It ran for 1600 miles up the length of the subcontinent; and on the inch-perfect accuracy of its plotted locations all other surveys and locations depended.

Clearly, too, the Great Arc was something rather special. The mathematical equations involved in its computation seemed to fill enough volumes to line a library, while the instruments used to gather the raw data were man-size contraptions of cast iron and well-buffed brass which were still lovingly displayed in the Survey of India's offices.

I ferreted further and I read on. Evidently the Great Arc was way ahead of its time. No scientific undertaking on such a massive scale had previously been attempted. Outside the regulated confines of Bourbon France and Georgian Britain, it was also the most minutely accurate land measurement on record. The accuracy was all-important because the Arc had as much to do with physics as with surveying. It was not simply an attempt to measure a subcontinent but also, incredibly, to measure and compute the precise curvature of the globe.

More prosaically, it was conceived by an elusive genius called William Lambton and was inspired by the first British conquests in the extreme south of India. The year was 1800. In North America Meriwether Lewis and William Clark had yet to set out on their epic journey across the continent. Australia was barely a penal colony; Napoleon still held Egypt; and most of the rest of Africa remained shrouded in that mysterious 'darkness' which was simply Europe's ignorance.

Only in Asia, and especially India, was an embryonic imperialism detectable. Already the infant was outgrowing the womb of trade, stretching and kicking prodigiously, and taking its first unsteady strides towards dominion. Just so the Great Arc. Mirroring the progress of empire, it would forge tentatively and then inexorably inland. During forty years of high-risk travel, ingenious improvisation and awesome dedication it would come to embrace the entire length of India. And when Lambton, its endearing founder, died in central India, it would be carried to its grand Himalayan finale by the bewhiskered and cantankerous martinet who was Colonel George Everest.

At the time the scientific world was frankly amazed. The Arc was hailed as 'one of the most stupendous works in the whole history of science'. It was 'as near perfect a thing of its kind as has ever been undertaken'. Lambton and Everest 'had done more for the advancement of general science than ... any other body of military men'. Their celebrity was assured. Lambton was internationally fêted. George Everest was knighted by Queen Victoria; and in his honour that peak, whose discovery and measurement the Arc had made possible, was duly named.

Yet today they are utterly forgotten. Lambton is not even among the fifteen thousand worthies included in *Chambers' Biographical Dictionary*. Everest is just a mountain. The progress of nineteenth-century invention was such that their science was almost instantly superseded. It now features only in histories of cartography, in academic critiques of imperialism, and in the dusty records of the Indian Survey. The Great Arc, like the great auk, has been consigned to oblivion.

Which is a pity, for it deserves better. At a time when to foreigners India was more a concept than a country, a place of uncertain extent and only fanciful maps, the Great Arc and the surveys based on it were indeed tools of imperial dominion as well as scientific enterprises. But thanks to that voluminous documentation and to George Everest's published memoirs, the story of the Great Arc transcends both its science and its politics. Uniquely for an official scientific venture, we can savour the setbacks, share the excitement, discern the personalities. In writing this book the challenge has not been that of embroidering bare facts with vivid shades of plausible detail but of stitching into the riot of authentic adventure a thread of scientific and political plausibility. Given half a chance, jungle mishaps would have put paid to the science, personal vendettas would have obliterated the politics, and the tigers would have made off with the narrative.

If the impression given is less that of a scientific set-piece and more of a monumental example of human endeavour, then so it was. Travelling India with an eye on the Arc, I found it impossible not to become obsessed by the sheer audacity of the enterprise. Like Mount Everest, which seen from afar looks a respectable peak but not obviously the world's highest, so the Arc viewed from a distance of two hundred years looks impressive but slightly quixotic. Get up close, though, breathe the sharp air and sense the monstrous presumption, and the Arc like the mountain soars imperiously to dwarf all else. Measuring the one, like climbing the other, is revealed as the ultimate challenge of its age.

Know . . .
That on the summit whither thou art bound
A geographic Labourer pitched his tent,
With books supplied and instruments of art,
To measure height and distance; lonely task,
Week after week pursued!

From 'Written with a slate pencil on a stone
on the side of the mountain of Black Comb',
WILLIAM WORDSWORTH, 1818

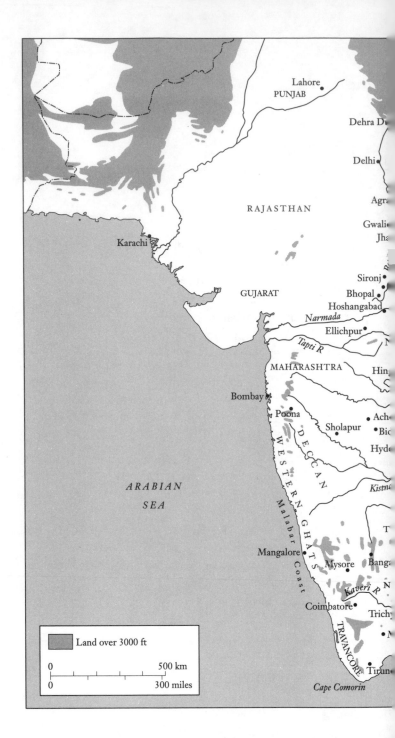

Lahore
PUNJAB

Dehra D

Delhi

Agra

RAJASTHAN

Gwali
Jha

Karachi

Sironj
Bhopal
Hoshangabad

GUJARAT

Narmada

Ellichpur

Tapti R

MAHARASHTRA

Hin

Bombay

Poona

Sholapur

Ach
Bic

Hyde

ARABIAN
SEA

DECCAN

Kistna

WESTERN GHATS

Malabar Coast

Mangalore

Mysore

Banga

Kaveri R

Coimbatore

Trich

TRAVANCORE

Tirun

Land over 3000 ft

0   500 km
0   300 miles

Cape Comorin

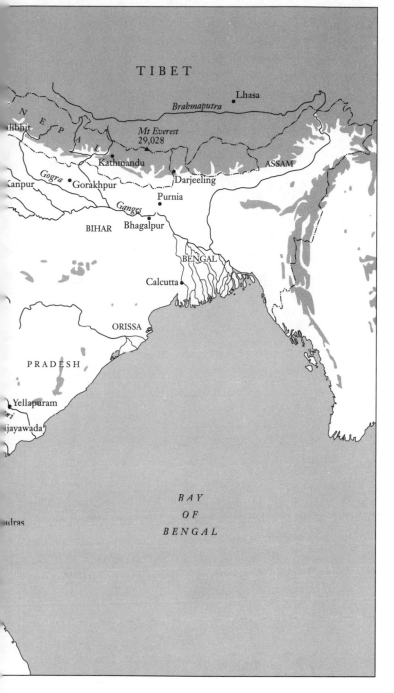

India

ONE

A Baptism of Fever

The word 'jungle' comes from India. In its Hindi form of *jangal*, it denotes any area of uncultivated land. Indian jungles are not necessarily forested, and today less so than ever. But well away from centres of population there do still survive a few extensive and well-wooded jungle tracts, especially in eastern and central India. Often they are classed as game sanctuaries, a designation which implies few facilities for the visitor but some much-advertised protection for the wildlife.

Here tigers and elephants yet roam, hornbills flap about in the canopy like clumsy pterodactyls, and hump-backed boar rootle aggressively through the leaf mould. In the dry season a safari might seem an attractive prospect. But be warned: 'dry' is high-baked. Like splintering glass, dead leaves explode underfoot to alert the animals. The tracks of crumbled dirt are hard to follow, spiked with ferocious thorns, and spanned by man-size webs patrolled by bird-size spiders.

The wet season is worse still. Then, the vegetation erupts. The tracks become impassable, and the air fills with insects. Only fugitives take to the jungle in the monsoon. Fugitives and, in days gone by when maps were rare, surveyors. In the year 1819, in just such a tract between the Godavari and Kistna rivers in what is now the south-eastern state of Andhra Pradesh, an English Lieutenant, lately attached to the Great Trigonometrical Survey of India and uncommonly keen to make his mark, underwent a baptism of fever.

Matters had gone badly for the twenty-eight-year-old Lieutenant from the start. Barely a month into this, his first season in the field, he had been confronted by a mutiny. 'The infliction of corporal punishment is an odious task,' he noted. But it was either that or abandoning the assignment. His escort obviously knew the perils of the monsoonal jungle and had seized every chance of escaping from the camp back to the city of Hyderabad. Something had to be done. Not without misgivings, the Lieutenant ordered one of these defaulters to be thrashed, whereupon the whole troop, about forty in number, took up their weapons and announced that they would decamp *en masse*. The British bluff had been called; in this insignificant and still today unfashionable corner of the subcontinent the myth of empire was at stake.

As might be inferred, by 1819 the British were already well on their way to becoming masters of India. Some areas had been won by conquest and were now under direct British rule; others were merely attached by treaty and remained nominally independent states under their own rulers. This was the case with the large principality of Hyderabad, through whose densest jungle the Kistna and Godavari rivers converged on the coast. Special permission had been obtained for the Great Trigonometrical Survey to operate in Hyderabad; but in 'a native state' the standards of subservience exacted in areas under direct British rule could not be taken for granted.

In fact, they could seemingly not be taken at all other than at the point of a gun. The mutineers, who now repaired to the nearby shade of a mango orchard, comprised a detachment of local troops lent by Nizam Sikander Jah of Hyderabad to protect and assist the British survey. In addition, the Survey had its own escort of twelve men who had been recruited in British territory, were paid out of the Survey's budget and had already amassed many years of loyal service. This in-house escort was now ordered to load muskets and take aim at the mutineers. A volley into their midst was threatened if they did not immediately surrender.

The ploy worked. The mutineers submitted, and this time the Lieutenant offered no apology for calling for the cane. Three men were publicly flogged, then dismissed; and thus, the Lieutenant tells us, 'was settled, very early in my career, a disputed point which had been a source of constant contention and annoyance to Colonel Lambton ever since his entering into the Nizam's territory'.

Colonel Lambton was the originator and now Superintendent of the Great Trigonometrical Survey of India. For seventeen years he had been spinning a web of giant geometry across the Indian peninsula without ever having had to thrash any of its teeming peoples. Tactful, patient and indestructible, Lambton seemed immune to India's frustrations, the result of a long wilderness experience in North America and of an attachment to science so obsessive and disinterested that even his critics were inclined to indulge him. Colonel Lambton beguiled India; but Lieutenant George Everest, his eager new assistant, chastised it.

The name, incidentally, was pronounced not 'Ever-rest' (like 'cleverest'), but 'Eve-rest' (like 'cleave-rest'). That was how the family always pronounced it, and the Lieutenant would not have thanked you for getting it wrong. Years later a fellow officer would make the mistake of calling him a 'Kumpass Wala'. No offence was meant. 'Kumpass Wala', or 'compass-wallah', was an accepted Anglo-Indian term for a surveyor. Everest, however, accepted nothing of the sort. He detested what he called 'nicknames' and, though it was not perhaps worth a dawn challenge, he demanded – and received – abject apologies. Getting on the wrong side of George Everest was an occupational hazard with which even British India would only slowly come to terms.

With the mutiny quelled and the mutineers 'finding that, when they knew me better, good behaviour was a perfect security against all unkindness', a self-righteous Everest pressed on for the jungles beside the Kistna. It was July, the month when the monsoon breaks. On time, the heavens duly opened just as he climbed a hill to his first observation post.

Survey work was conducted during and immediately after the monsoon because, regardless of the discomfort, it was only then that the dust was laid and the heat-haze dispersed. In the interludes of bright sunshine, the atmosphere was at its clearest; in fact it became so transparent that Everest fancied he could see forever and that 'the proximity of objects was only to be judged by their apparent magnitudes'. Trigonometrical surveying depended on the sighting of slender signal posts over distances of more than twenty miles. The monsoon's perfect visibility was therefore ideal. Spying a long dark ridge all of sixty miles to the east, Everest despatched his four best signalmen to occupy its heights. The ridge, he understood, was called Panch Pandol, and the signalmen were to erect their flagpole there in readiness for his observations. Meanwhile he continued south to the Kistna with the rest of his party.

Although visibility was greatly enhanced by the monsoon, mobility was not. Dry riverbeds instantly became raging torrents full of uprooted trees. The Musi, a tributary of the Kistna, rose so rapidly that Everest found himself cut off from his supplies. On iron rations therefore, and denied the normal crossing on the Kistna, which was on the other side of its confluence with the Musi, he headed downstream to where an alternative ferry was said to operate at a spot about fifty miles above the modern city of Vijayawada.

The Kistna, one of India's mightiest rivers, was now thrashing dementedly over steeply shelving rock like a panic-stricken patient beneath the surgeon's knife. Crossing it meant trusting oneself to a coracle, a small circular vessel, more bowl than boat, made of woven rattans and faced with hide. Everest likened it to a leather basket. Such craft, still used in many parts of India, are highly portable and sometimes formed part of a surveyor's outfit. Although not so provided, Everest found one abandoned by the river.

While it was undergoing the necessary repairs at the hands of the village cobbler, Everest ordered his 'carriage-cattle' to be swum across the flood. Fortunately they were not actually cattle, ox-carts being useless in roadless jungle, but a species he deemed 'more at home in the water than any other quadruped', namely elephants. As he also noted, elephants are extraordinarily sagacious. The Survey's beasts duly swayed to the bank, took a long look at the rocks and the raging waters, assessed the mix of caresses and curses on offer, and opted to stay dry. 'Probably it was fortunate,' Everest adds, 'for these powerful animals . . . are, from the size of their limbs, in need of what sailors term sea-room, and in a river like the Kistna . . . were very liable to receive some serious injury.'

This reverse meant a change of plan. Dr Henry Voysey, one of Everest's two British companions and the Survey's geologist-cum-physician, was left on the north bank with the main party plus elephants, horses, tents and baggage. Meanwhile Everest and a dozen men, balancing the Survey's cumbersome theodolite between them, crossed to the other side. Three trips had to be made; and since the coracle had to undergo repairs after each, it took most of the day. Then, deceived by the visibility into thinking it was only a couple of miles away, Everest immediately set out for his next observation post.

The couple of miles turned out to be twelve. They included both jungle work and rock-climbing. By the time the hill of Sarangapalle was reached it was dusk, and big black clouds, aflicker with lightning, were piling up overhead. 'At last,' noted Everest, 'when all their batteries were in order, a tremendous crash of thunder burst forth, and, as if all heaven were converted into one vast shower-bath, the vertical rain poured down in large round drops upon the devoted spot of Sarangapullee.'

Tentless in the deluge, Everest and his men bent branches

to make bivouacs. His own was improved by a bedstead and an umbrella, between which he slept the sleep of the utterly exhausted, oblivious alike of his squelching tweeds, puddled bedding and benighted followers. 'These evils might have been borne without any ill effects,' he insists, 'but for other circumstances of more serious consequence.'

The natives of India, according to Everest, 'with their minds bowed down under the incubus of superstition', attributed all fevers to witchcraft, and ignored natural causes. He, on the other hand, while amused by the idleness and absurdity of these doctrines, knew better. Malaria and 'typhus' fevers were alike the result of 'a poisonous influence in the air' which emanated from moist and 'unwholesome' soils. Under the impression that he was contributing to medical research, he examined the different schists and shales, the crystalline sandstone of Sarangapalle, the blue limestones of the Kistna and the porous sandstone of the Godavari in minute detail. These, he believed, were the 'other circumstances' which would prove of such serious consequence for his survey.

At the time most of his contemporaries shared these medical views. But, in a nice case of geographical coincidence, Hyderabad would host a further attempt to discover the natural causes of malaria. Seventy years later in a house in Begampet, now a suburb of Hyderabad city, Surgeon Ronald Ross would experiment on the insects of the Kistna-Godavari jungles and trace the malaria parasite to the anopheles mosquito. Everest's ideas of 'malarial vapours' would thereby be exposed as every bit as idle and absurd as those of his followers.

After observing from Sarangapalle, he recrossed the Kistna and rejoined his camp to head north towards the Godavari. On the way he conducted observations to prominent hills like that to which he had earlier sent his signalmen. The survey on which he was engaged was what was known as a 'secondary triangulation'. It was intended to cover all the country between the Kistna and the Godavari with a network of imaginary triangles whose sides connected intervisible observation posts.

Triangulation means simply 'triangle-ing', or conceiving three mutually visible reference points, usually on prominent hills or buildings, as the corners of a triangle. Knowing the exact distance between two of these points, and then measuring at each the angles made by their connecting sight-line with those to the third point, the distance and position of the third point can be established by trigonometry. One of the newly determined sides of this triangle then becomes the base for a second triangle embracing a new reference point whose position is determined in the same way. Another triangle is thus completed and one of its sides becomes the base for a third, And so on. A web, or chain, of triangles results; and Everest's job was to extend this web of triangulation over the whole Kistna-Godavari region.

The positions of these vital reference points could have been established by careful observation of the stars. But as Everest would repeatedly emphasise, astronomical observations only gave the desired degree of accuracy if conducted over many months, preferably years, from well-equipped and professionally-manned observatories. Constructing and operating such observatories across a subcontinent was out of the question; and for reasons that were only partly understood at the time, observatories seemed to be affected by their surroundings. Better and simpler was the geometrical approach of triangulation. It was not quicker. The Great Trigonometrical Survey had taken the field twenty years before Everest became involved and would not complete its work until twenty years after he left. Nor was it necessarily cheaper. As Everest was about to discover, in terms of lives lost and rupees spent the cost would exceed that of many contemporary Indian wars. But triangulation was well-tried, accurate to the point of mathematical certainty, and so more acceptable to the scientific world.

Such a survey still depended on the occasional astronomical

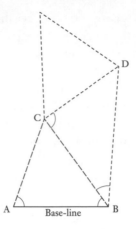

Fig 1   BASIC TRIANGULATION

Given the distance between points A and B (the base-line), the distance
from each to point C is calculated by trigonometry using the AB base-line
measurement plus the angles CAB and ABC as measured by sighting with
a theodolite. The distance between C and B having now been established,
it may be used as the base-line for another triangle to plot the position of
D. CD may then be used as the base for a third triangle, and so on.

observation in order to locate and orientate its triangles in
terms of the earth's grid of latitude and longitude. It also
depended on the occasional measurement along the ground.
Known as a 'base-line', this was needed to get the triangulation
going in the first place by establishing the distance between
the first two points. It was also a useful way of verifying the
accuracy of a protracted triangulation, since the distance
between any two points as established by triangulation could
be checked by another actual measurement on the ground.

In the case of a 'secondary triangulation' like Everest's
between the Kistna and Godavari, base-line measurements and
astronomical observations were not necessary. Everest's job
was to connect trig stations along the east coast (whose relative
positions had already been established by the more elaborate

methods and instruments of primary triangulation) with those of another chain of even more exacting primary triangulation about a hundred miles inland to the west.

The latter roughly followed the 78-degree meridian (or north–south line of longitude) and consisted of a continuous chain of triangles which had been carried from Cape Comorin at the tip of the Indian peninsula as far north as Hyderabad, a distance of about seven hundred miles. Already this 'series' was known as the Great Arc of the Meridian. As well as providing the spine on which the whole skeleton of the Great Trigonometrical Survey depended, it was the aspect of the Survey's work which most appealed to George Everest. In fact his present assignment he saw mainly as a way of proving that he was pre-eminently qualified to succeed Colonel Lambton as the grand master of the Great Arc.

To one like Everest who happened to have been baptised (and so probably born) in the London parish of Greenwich, meridians must early have meant something. Greenwich had been the site of England's Royal Observatory since the seventeenth century. British navigators and surveyors regarded the Greenwich meridian, or 'mid-day' line (because at any point along a north–south meridian the sun reaches its zenith at the same time), as the zero from which they calculated all longitudinal distances and from which on maps and charts they extended the graticule, or grid, of the globe's 360 degrees of longitude. Later in the nineteenth century this British convention would win international approval. Greenwich Mean Time would become established as a world standard and the Greenwich meridian would be universally recognised as 0 degrees longitude. It became, in fact, the north–south equivalent of the east–west equator at 0 degrees latitude.

Everest therefore knew about meridians from childhood and may well have been intrigued by the problems of determining them. Later, at the Royal Military Academy at Woolwich, he had studied the mathematics, mechanics and measuring techniques essential for an officer joining the artillery. But his family background was not scientific, his father being a solicitor; and apart from some basic survey work in Java when British forces had invaded and occupied that island during the Napoleonic wars, his career had thus far differed little from that of other army officers in India. Appointment to the Great Trigonometrical Survey was his big opportunity. Neither mutiny, flood nor fever was going to impede his determination to excel.

Returning from the soaking at Sarangapalle, Everest revisited his first observation post, erected his theodolite – the instrument used for measuring the angles between sight-lines – and scanned the distant ridge of Panch Pandol through its telescope. Nothing had been heard of his signalmen for three weeks; nor was there now any sign of their signals. But a few days later a gap was noticed in the dark vegetation which covered the ridge. Day by day it was seen to grow into two sizeable clearings. 'After a fortnight's further waiting I had sufficient daylight behind [the clearings] to distinguish the colours of the Great Trigonometrical Survey flying on the one spot and a signal-marker on the other.' Bearings could now be taken to ascertain the angle between the sight-line to this new marker and that to another marker at an already established observation post.

Measuring such angles was the essence of trigonometrical survey work. Another triangle was thus completed and, once the sight-line to Panch Pandol had been calculated, it could serve as the base for the next triangle. The whole party then moved on towards the ridge to begin their observations anew and, in the case of the impatient Everest, to seek some explanation for his signalmen's unconscionable delay in reaching Panch Pandol.

The explanation was soon obvious. Almost immediately the trail plunged into the formidable jungle region which now comprises the Pakhal and Eturnagaram game sanctuaries. The forests were of ebony and teak, and the trees 'seventy, eighty, and even ninety feet high, thickly set with underwood, and infested with large tigers and boa constrictors'. As the Survey gingerly hacked its way forward, Everest began to think more kindly of his signalmen. 'How . . . without water or provisions, and with the jungle fever staring them in the face, they could have wandered through such a wilderness until they selected the most commanding points for a station, utterly, I confess, surpasses my comprehension.' His comprehension would soon again be found wanting. The scene which greeted him on arrival was even more impressive.

> When I saw the dreadful wilderness by which I was surrounded; when I saw how, by means of conciliating treatment and prompt payment, my people had managed to collect a sufficient body of hatchet-men to clear away every tree which in the least obstructed the horizon over a surface of nearly a square mile; and when [I saw how] the gigantic branches of these were cut off and cleared away leaving only the trunks as trophies, – then – then I learned to appreciate the excellent management of Colonel Lambton who had been enabled to train up so faithful a body of men.

Then, somewhat incidentally, he also 'learned how to value the natives of southern India'. But it was a lesson that was easily forgotten. Giving credit to subordinates would not come naturally to George Everest. From Panch Pandol he despatched his advance party to a hill site even deeper in the jungle and near the banks of the Godavari. Again the days slipped by with no sign of them; again Everest fretted and fumed. He sent out a second party to look for them, then a third. Finally he despatched his chief sub-assistant Joseph Olliver, who with Dr Voysey made up his entire British staff.

Olliver eventually reached the hill and hoisted the flag; but his news was not good. Most of the previous signalmen had succumbed to fever; some were near death. Should the whole survey party proceed to Yellapuram (the village after which the new site was named) the risks would be immense. Everest was unimpressed. Desperate to complete his assignment and so win the approval of Colonel Lambton, he reckoned that all risks were warranted.

The trail from Panch Pandol to Yellapuram wound through 'the wildest and thickest forest that I had ever invaded'. It took three days; but at least the weather stayed fine and the vegetation was at its most spectacular after the recent rains. Voysey and Everest rejoiced as they rode, then quipped as they climbed. At last the canopy thinned and, seeing again the sky and the summit, both men spontaneously roared a favourite Shakespearian couplet:

> Night's candles are burnt out, and jocund day
> Stands tiptoe on the misty mountain's top.

Everest, however, misquoted; and neither man seems to have been aware of Romeo's next and more cautionary line: 'I must be gone and live, or stay and die.'

After they dismounted at Yellapuram, the oppressive silence of the jungle brought to Everest's mind a wilderness scene from the *Arabian Nights*. There was a spectacular view up the Godavari and, beside and beyond it, three excellent heights from which to complete his survey. Congratulating himself that 'the end of my toilsome and laborious task seemed now to be within my grasp', he immediately sent out flag parties.

But no sooner had jocund day forsaken the misty mountain's top than fever struck. That evening Everest went down with what he called a violent typhus, the result of 'my day's ride through a powerful sun and over a soil teeming with vapour and malaria'. Dr Voysey succumbed soon after. Within five days most of their followers, including escort, signalmen, porters, mahouts and runners, nearly 150 in all, were also prostrated.

It seemed indeed as if at last the genius of the jungle had risen in his wrath to chastise the hardihood of those men who had dared to violate the sanctity of his chosen haunt. All hope of completing the work this season being now at an end, it remained only to proceed with as much expedition as possible towards Hyderabad . . . [and] to return, baffled and crippled, through an uninterrupted distance of nearly two hundred miles.

Dr Voysey took to his palanquin. Everest, lacking such a conveyance, had a stretcher made. For porters they looked not to their prostrate followers but to the retinue of 'a rebellious chief who aided my progress most manfully'. It took three weeks for them to reach Hyderabad, throughout which time 'the jungle fever pursued my party like a nest of irritated bees'.

When news of the disaster reached the city, all available carts, palanquins, elephants and camels were commandeered and sent out to bring home the sick. Most were indeed retrieved but, out of the total of 150, fifteen had died on the road and not one had escaped unscathed. The survivors, wrote a shaken Everest, 'bore little resemblance to human beings, but seemed like a crowd of corpses recently torn from the grave'.

So ended Lieutenant George Everest's first season in the employ of the Great Trigonometrical Survey. A long convalescence was necessary; it was anyway October, by which month the visibility had lost its champagne clarity. For Everest the experience had been an eye-opener. He recalled it with a mixture of horror and naivety which is seldom found in his other writings. It was not exceptional; greater catastrophes would overtake the Survey and many more lives would be lost. But it was a testing induction for a novice, and it was an ominous overture to an illustrious but controversial career.

Dr Voysey would never fully recover. Though he soldiered

on, he would die four years later from a recurrence of the Yellapuram malaria. Everest, too, would never regain what he calls 'the full vigour of youth'. In the following year he returned to Yellapuram to complete his observations but again succumbed to a 'violent attack of jungle fever'. The work was in fact completed by his dependable assistant Joseph Olliver. Meanwhile Everest, 'deeming it unwise to sacrifice myself for an unimportant object', took a year's sick leave and sailed to the Cape of Good Hope to convalesce. He would return to duty in 1822 but within a year was racked by fevers both old and new. Gruesome complications ensued which would temporarily reduce him to a cripple. In 1825, aged thirty-five, he would again sail away on sick leave, this time to England. He would not return to India for five years.

Critical for Everest, the period from 1820 to 1830 would prove even more critical for what he proclaimed to officials in London to be 'the greatest scientific undertaking of the kind that has ever been attempted'. By this he meant not the ambitious map-making programme of the Survey of India, nor even the rigorous methods of its Great Trigonometrical Survey, but the latter's supreme expression, the Great Indian Arc of the Meridian.

As his birthplace of Greenwich was to meridians, so George Everest would become to the Arc. The two became inseparable. The Arc would be his life's work, his dearest attachment, his near-fatal indulgence; and while he lived, his name would be synonymous with it. Yet it was not his brain-child, nor in large part his achievement – those honours belong to the less articulate genius of William Lambton. Nor, when Everest died, would he long be remembered for the Arc. Instead, his name was purloined for a peak.

It was not in his nature to decline the lasting fame of having his name 'placed a little nearer the stars than that of any other'. Even the controversy which the naming of Mount Everest would prompt is in character. On the other hand, his truculent spirit must surely be turning in its grave at being remembered only for the mountain and not for the measurement. Other than as convenient trig stations, mountains barely featured in his life. He saw the Himalayas only towards the end of his career and he hailed them then only as a fitting conclusion to the Great Arc. There is nothing to suggest that he was particularly curious as to their height.

Yet there was a connection between the Arc and the Himalayas, and there was a logic in naming the earth's greatest protuberance for Everest. For the Great Arc would solve the mystery of the mountains. The painstaking measurement of a meridian up through India's burning immensity would make possible the measurement of the ice-capped Himalayas. This is the story of both, of the Arc and of the mountains.

TWO

The Elusive Lambton

Everest's predecessor as Superintendent of the Great Trigonometrical Survey is less obviously commemor-ated. In fact, to this mild and reclusive man of science there seems to be no memorial at all. There is not even any structure which can certainly be associated with his work. It has, though, been my privilege to stand at his graveside. The place proved hard to find and was not at all distinguished. I doubt if anyone has been to Hinganghat to look for it in the past fifty years. The locals knew nothing of its whereabouts nor, until my wife began spelling out his epitaph, had they ever heard the name of William Lambton.

Luckily the day was a Sunday, for to our visit coinciding with morning mass in Hinganghat we owed the discovery. Enquiries about a Christian cemetery had at first been received with blank stares from the congregation of Keralan immigrants as they spilled forth into the fields. Then, with the organ still playing, there emerged a man of more bracing faith. Mr K.J. Sebastian, an English teacher, might rather have devoted his day of rest to his young family; but grasping the gist of my story, he leapt to the challenge and sped off on his scooter, we following close behind, to explore the byways of the parish.

Hinganghat lies about fifty miles south of Nagpur and is as near the dead centre of India as anywhere. It also epitomises much that is unlovely about the country. Unless your business is cotton there can be no possible reason for turning off the Wardha road. Two large mills, their machinery housed in untidy hangars of rusty corrugated sheeting, dominate the prairie landscape and provide some badly paid employment. The rhythm of their shifts regulates Hinganghat's day, and to the farmyard ordure of what is otherwise just an overgrown village they add an oily slick of industrial squalor. As the driver had warned, 'Hinganghat like shit.'

Behind a street frontage of tented tea-stalls and tyre-repair shops a game of cricket was being played on a piece of waste land. It was our third point of call. Dodging the worm castings of human excrement which dotted the pitch like daisies, we trailed round the outfield towards a small whitewashed mosque. According to a report of 1929 Lambton's grave had been joined by others and the spot consecrated as a Christian cemetery. Since no such place now existed in Hinganghat's collective memory, and since, apart from Christians, only Muslims bury their dead, Mr Sebastian thought that the Maulvi, the local prayer-leader, might be able to help. Yes, said the Maulvi, there had been Christian tombs in what he called the Muslim cemetery, and although the hallowed ground had lately been built on by squatters, two were still intact.

One, mysteriously known as 'the Belgian's Stone', turned out to be an obelisk within a circular walled enclosure which now served the squatter colony as a central urinal. The other was just a plain oblong plinth with the raised outline of a casket on its surface. Children used it for climbing on. There was no headstone and the whole sepulchre had at some point been encased in mortar. Into this mortar, when wet, someone had written three lines of text with a finger. The letters were ill-formed and were much too large ever to have conveyed more than the most basic information. There might originally have been twenty, and they looked to have been copied, perhaps from an earlier inscription, by someone not confident with Roman script. The mortar was now crumbling so badly that barely half were legible. But the 'L', 'A', 'M' and 'B' running along the top line were still clear. So was the word 'DATE', an annoyingly superfluous survival. It was followed by the three numerals '1', '7' and '6', at which point the mortar had broken away.

If this date was to be read as seventeen-sixty-something, it was wrong. There could hardly be any question that this was indeed Lambton's resting place, but he died in 1823. Moreover, seventeen-sixty-anything was rather early for a European grave in such an out-of-the-way place. It occurred to me, therefore, that it must be a birth date. On slender evidence Lambton's birth is usually given as 1753. This would make him fifty when he started on the Great Arc, sixty-six when Everest joined him in Hyderabad, and an impressive but improbable seventy when he died. He was still in the field at the time, indeed looking forward to carrying his triangles on to Agra in the north of India, another two years' work at least. Amongst Europeans exposed to India's lethal climate seventy-year-olds were as rare then as centenarians today. A working seventy-year-old would have been a great curiosity and would certainly have attracted much contemporary comment. On the whole, then, I was ready to give the tomb the benefit of the doubt. Sometime in the early 1760s seemed a more plausible birth date than 1753. It also disposed of a decade-long void when Lambton, supposedly in his twenties, unaccountably disappears from the record.

Where he was born is more certain. It was on a debt-ridden farm in the North Riding of Yorkshire whose plight would oblige him to make the support of his impoverished parents an important career consideration. Early promise in mathematics won him a place in a grammar school and, in 1781, an Ensignship in an infantry regiment. With the 33rd Foot he promptly sailed for the war (of Independence) in America and was there promptly taken prisoner at York Town. After release he was ordered to the then wilderness of New Brunswick on the north-eastern seaboard. He helped divide and apportion its land amongst British loyalists displaced by the American victory, and was involved in surveying and delineating what now became the boundary between British Canada and the United States.

Nine years later, apparently as a result of an oversight, he was still in New Brunswick and still an Ensign, although drawing additional pay as a civilian Barrack Master. A hint, however, that his years in the wilderness were numbered came in 1793 when he was unexpectedly promoted; 'to his astonishment,' in the words of the *Royal Military Calendar*, 'he found himself a Lieutenant.' Two years later he was ordered to choose between the army and his civil appointment; and having plumped for the army, in 1796 he was posted to India.

The man behind this flurry of orders was the new Commandant of Lambton's regiment, a twenty-seven-year-old Colonel called the Honourable Arthur Wesley. Wesley, better known by the later spelling of 'Wellesley', would one day become better known still as the Duke of Wellington, victor of Waterloo. Besides commanding the 33rd Foot, he was the younger brother of Richard Wesley (or Wellesley), then Earl of Mornington and also about to leave for India. Richard had been appointed Governor-General of the British possessions in the East and blithely perceived his task as that of augmenting them. Young Arthur and his regiment, including the elusive Lambton, were in for a busy time.

The two men first came face to face when sailing on the same ship from Calcutta to Madras in 1798. Arthur Wellesley, en route to a war which his brother was aggressively fomenting with the ruler of the independent state of Mysore, was much too preoccupied to quiz the newcomer. He was, though, puzzled by him. Lambton, now perhaps in his late thirties, had obviously been out of circulation far too long. Tall, strongly built and clean-shaven, with reddish hair already thinning, he was awkward in society and unusually economical in his habits. '[His] simplicity of manner gave many people a very inadequate idea of his powers of mind and knowledge of the world,' recalled John Warren, an old friend. 'Some peculiarity of manner adhered to him from having lived so long out of the world. His face wanted expression, and the old accident gave a cast to his eye.' The 'old accident' had occurred while observing a solar eclipse in Canada. Omitting the elementary precaution of attaching a smoked glass to his telescope, Lambton had partially lost the use of his left eye. The result was a slightly glazed expression and a heightened concern for any subordinate using such instruments under his direction.

Despite these peculiarities, Arthur Wellesley was impressed by Lambton's abilities. He asked others to corroborate them and, when their ship reached Madras, he invited Lambton to share his residence. Whatever thirteen years in the wilderness had done to the man's social skills, they had not been wasted professionally. Lambton had somehow acquired a familiarity with higher mathematics, mechanics and astronomy which would have been impressive in London, let alone India. On arrival in Calcutta he had contributed a paper, full of the most awesome mathematical equations, to *Asiatick Researches*, India's leading academic publication. Invitingly titled 'Observations on the Theory of Walls', it demonstrated that for any fortifying wall there was an optimum depth of foundation which it was mathematically pointless to exceed. Such knowledge, although of limited use at a time when the British in India had taken the offensive, convinced Colonel Wellesley that Lambton was far from being the dolt he appeared. Lambton continued to regret that the Colonel never spoke to him. Perhaps Wellesley was anxious not to betray his scientific ignorance. But clearly he valued Lambton's company and would soon prove a useful patron.

Lambton's opportunity came courtesy of the war with Mysore which finally got underway in 1799. At the time the British had been established at Madras for more than 150 years. Merchants of the English East India Company had been buying cotton textiles from this part of peninsular India since the early seventeenth century and took great pride in the fort, and now city, which they had founded at Madraspatnam in 1640. But it was not until a century later, when wars in Europe had embroiled them with their French rivals based at nearby Pondicherry, that the British had begun to take an interest in Indian territory as opposed to trade. By then there were numerous other British, or rather East India Company, trading settlements around the coasts of India, and it was in fact from one of these, Calcutta, that the first move towards an Indian dominion had been made.

Between 1756 and 1766 Company men in Calcutta deployed troops intended for another war with their French rivals to overthrow the local Nawab and establish a claim to the revenues of Bengal. One of the largest and richest provinces in all India, Bengal comprised the modern Bangladesh plus the neighbouring Indian states of West Bengal, Bihar and Orissa. It was from northern Bihar's border with Nepal that British officials first glimpsed the sawtooth profile of the high Himalayas, and it was from this substantial Bengal bridgehead that British forces in northern India would begin their inexorable march up the Gangetic plain towards the old Mughal capital of Delhi.

Meanwhile Madras in the south and Bombay in the west had remained separately governed 'Presidencies' (because each had its own British 'President', or Governor). Still dedicated to the ancient imperatives of trade, they were much more vulnerable to attack than Bengal, whose officials increasingly regarded them as political liabilities, a feeling which was intensified when in the 1770s Calcutta was named the capital of British India and its Governor was appointed Governor-General over all the British holdings in India.

At the time Madras, although relieved of the French challenge from Pondicherry, confronted an Indian challenge from the expansive ambitions of an upcountry neighbour in the state of Mysore, roughly the modern Karnataka. There ensued no fewer than four Anglo–Mysore Wars, that of the Wellesleys and Lambton being the Fourth. It was also much the most one-sided. The gauntlet first thrown down in the 1760s–80s by Mysore's Haidar Ali, a formidable campaigner, had come to look more like a glove-puppet when tossed into the ring in the 1790s by his quixotic son Tipu Sultan. By then the British, buoyed by their successes in Bengal, were capable of over-whelming any opposition and happily construed all but abject compliance as punishable defiance.

Tipu Sultan had counted on French support. To this end he had reversed the one-way traffic of colonial diplomacy by despatching an impressive mission to Versailles. It had arrived in France in 1788 only to find Louis XVI desperately trying to stave off his own crisis – the deluge which within a year would plunge France into Revolution. No Franco–Mysore alliance resulted, and in India Tipu now stood against the mighty concentration of British power. He remained defiant. Dubbed the 'Tiger of Mysore', he delighted in a working model, complete with sound effects, of a tiger devouring an English soldier (now in London's Victoria and Albert Museum). But in the Third Anglo–Mysore War of 1790 it was the tiger who was severely mauled; and in the Fourth of 1799 it remained only to despatch him.

Lambton played his part in this war with distinction. By consulting the stars he was able to avert a disaster when during a night march General Baird mistakenly led his column south towards enemy lines rather than north to safety; and at the great set-piece siege of Tipu's stronghold at Srirangapatnam he set a rather better example of derring-do than the future 'Iron Duke'. The war itself, waged with such overwhelming superiority, proved little more than the expected tiger-hunt. It lasted just four months. Srirangapatnam was ravaged with an ardour worthy of Attila the Hun, and Tipu was found slain amongst the ruins.

Rounding up the spoils took longer and was much more gratifying. The territories of Mysore stretched across peninsular India as far as the west, or Malabar, coast and south almost to its tip. Following Calcutta's example in Bengal, Madras had at last acquired a sizeable hinterland of Indian real estate, most of which would henceforth be directly ruled by the British.

It was while travelling with Arthur Wellesley and his staff across this fine upland country of teak woods and dry pasture, subduing a recalcitrant chief here and plundering a fortress there, that Lambton conceived his great idea.

As when New Brunswick was settled, the country was virtually unknown to the British. To define it, defend it and exploit it, maps were desperately needed, and two survey parties duly took the field in 1799–1800. One concentrated on amassing data about crops and commerce. Its three-volume report, a rambling classic of its kind, would include such gems as an account of cochineal farming – or rather ranching, for the small red spiders from which the dye is extracted required only tracking and culling as they spun their way along the hedgerows, multiplying prodigiously.

The other survey was a more formal affair, similar to surveys already undertaken in Bengal. It was equipped with theodolites for triangulation, with plane-tables for plotting the topographic detail, and with wheeled perambulators and steel chains for ground measurement. Colonel Colin Mackenzie, who conducted it, was another noted mathematician who had originally forsaken his home in the Hebridean Isle of Lewis to visit India in order to study the Hindu system of logarithms. His Mysore Survey was a model of accuracy and the maps which it yielded faithfully delineated the frontiers of the state as well as indicating 'the position of every town, fort, village . . . all the rivers and their courses, the roads, the lakes, tanks [reservoirs], defiles, mountains, and every remarkable object, feature, and property of the country'. Additionally, Mackenzie collected information on climate and soils, plants, minerals, peoples and antiquities. The last was his speciality. In the course of the Mysore Survey and other travels, he amassed the largest ever collection of Oriental manuscripts, coins, inscriptions and records. Congesting the archives of both India and Britain, the Mackenzie Collection was still being catalogued a hundred years later.

Under the circumstances, Lambton's big idea to launch yet a third survey looked like a case of overkill; and with Mackenzie's efforts promising to make Mysore the best-mapped tract in India, Lambton anticipated official resistance. But as Arthur Wellesley now appreciated, his subordinate was proposing not a map, more a measurement, an exercise not just in geography but in geodesy.

Geodesy is the study of the earth's shape, and it now appeared that while holed up through a dozen long Canadian winters Lambton had made it his speciality. Studying voraciously, reading and digesting all the leading scientific publications, he had taken a particular interest in the work of William Roy, founder of the British Ordnance Survey, and of Roy's even more distinguished mentors in France.

Surveying of a basic nature had been among Lambton's early responsibilities in Canada. Some old maps of New Brunswick actually show a 'Lambton's Mountain'. It is not very high and the name, unlike Everest's, would not stick. Instead it became 'Big Bald Mountain' – which was more or less what Lambton would also become. But such surveying, although based on the simple logic of triangulation, was child's play compared to what the Cassini family in France and William Roy in Scotland and England had been attempting.

Triangulation, together with all its equations and theorems (like that of Pythagoras), is strictly two-dimensional. It assumes that all measurements are being conducted on a plane, or level surface, be it a coastal delta or a sheet of paper. In practice, of course, all terrain includes hills and depressions. But these too can be trigonometrically deduced by considering the surface of the earth in cross-section and composing what are in effect vertical triangles. The angle of elevation between the horizontal and a sight-line to any elevated point can then be measured and, given the distance of the elevated point, its height may be calculated in much the same way as with the angles on a horizontal plane. Thus would all mountain heights be deduced, including eventually those of the Himalayas. Adding a third dimension was not in theory a problem.

However, a far greater complication arose from the fact that the earth, as well as being uneven, is round. This means that the angles of any triangle on its horizontal but rounded surface do not, as on a level plane, add up to 180 degrees. Instead they are slightly opened by the curvature and so come to something slightly more than 180 degrees. This difference is known as the spherical excess, and it has to be deducted from the angles measured before any conclusions can be drawn from them.

For a local survey of a few hundred square miles the discrepancies which were found to result from spherical excess scarcely mattered. They could anyway be approximately allocated throughout the measurement after careful observation of the actual latitude and longitude at the extremities of the survey. This was how Mackenzie operated. But such rough-and-ready reckoning was quite unsatisfactory for a survey of several thousand square miles (since any error would be rapidly compounded); and it was anathema to a survey with any pretensions to great accuracy.

The simplest solution, as proposed by geographers of the ancient world, was to work out a radius and circumference for the earth and deduce from them a standard correction for spherical excess which might then be applied throughout any triangulation. But here arose another and still greater problem. The earth, although round, had been found to be not perfectly round. Astronomers and surveyors in the seventeenth century had reluctantly come to accept that it was not a true sphere but an ellipsoid or spheroid, a 'sort-of sphere'. Exactly what sort of sphere, what shape of spheroid, was long a matter of dispute. Was it flatter at the sides, like an upright egg, or at the top, like a grapefruit? And how much flatter?

Happily, by Lambton's day the question of the egg versus the grapefruit had been resolved. In the 1730s two expeditions had been sent out from France, one to the equator in what is now Ecuador and the other to the Arctic Circle in Lapland. Each was to obtain the length of a degree of latitude by triangulating north and south from a carefully measured base-line so as to cover a short arc of about two hundred miles. Then, by plotting the exact positions of the arc's extremities by astronomical observations, it should be possible to obtain a value for one degree of latitude. Not without difficulty and delay – the equatorial expedition was gone for over nine years – this was done and the results compared. The length of a degree in Ecuador turned out to be over a kilometre shorter than that in Lapland, in fact just under 110 kilometres compared with just over 111. The parallels of latitude were thus closer together round the middle of the earth and further apart at its poles. The earth's surface must therefore be more curved at the equator and must be flatter at the poles. The grapefruit had won. The earth was shown to be what is called an 'oblate' spheroid.

There remained the question of just how much flatter the poles were, or of how oblate the spheroid was; and of whether this distortion was of a regular or consistent form. This was the challenge embraced by the French savants and by William Roy in the late eighteenth century. Instruments were becoming much more sophisticated and expectations of accuracy correspondingly higher. The pioneering series of triangles earlier measured down through France was extended south into Spain and the Balearic Islands and then north to link across the English Channel with Roy's triangles as they were extended up the spine of Britain. The resultant arc was much the longest yet measured and, despite a number of unexplained inconsistencies, provided a dependable basis for assessing the earth's curvature in northern latitudes, and so the spherical excess.

Lambton was now proposing to do the same thing in tropical latitudes, roughly midway between the equator and northern Europe. But like his counterparts in Europe, he played down the element of scientific research when promoting his scheme and stressed the practical value that would arise from 'ascertaining the correct positions of the principal geographical points [within Mysore] upon correct mathematical principles'. The precise width of the Indian peninsula would also be established, a point of some interest since it was now British, and his series of triangles might later be 'continued to an almost unlimited extent in every other direction'. Local surveys, like Mackenzie's, would be greatly accelerated if, instead of having to measure their own base-lines, they could simply adopt a side from one of Lambton's triangles. And into his framework of 'principal geographic points' existing surveys could be slotted and their often doubtful orientation in terms of latitude and longitude corrected. Like an architect, he would in effect be creating spaces which, indisputably sound in structure, true in form and correct in position, might be filled and furnished as others saw fit.

He could, however, scarcely forbear to mention that his programme would also fulfil another 'desideratum', one 'still more sublime' as he put it: namely to 'determine by actual measurement the magnitude and figure of the earth'. Precise knowledge of the length of a degree in the tropics would not be without practical value, especially to navigators whose charts would be greatly improved thereby. But Lambton was not thinking of sailors. As he tried to explain in long and convoluted sentences, his measurements aimed at 'an object of the utmost importance in the higher branches of mechanics and physical astronomy'. For besides the question of the curvature of the earth, doubts had surfaced about its composition and, in particular, the effect this might be having on plumb lines. Plumb lines indicated the vertical, just as spirit levels did the horizontal, from which angles of elevation were measured both in astronomy (when observing for latitude and longitude) and in terrestrial surveying (when measuring heights). But inconsistencies noted in the measurement of the European arc had suggested that plumb lines did not always point to the exact centre of the earth. They sometimes seemed to be deflected, perhaps by the 'attraction' of nearby hills. If the vertical was variable – as indeed it is – it was vital to know why, where, and by how much. New meridional measurements in hitherto unmeasured latitudes might, hoped Lambton, provide the answers.

Whether, reading all this, anyone in India had the faintest idea what Lambton was on about must be doubtful. But Arthur Wellesley warmly commended his friend's scientific distinction, Mackenzie strongly urged the idea of a survey which would surely verify his own, and Governor-General Richard Wellesley was not averse to a scheme which, while illustrating his recent conquests, might promote the need for more. The beauty of map-making as an instrument of policy was already well understood; it would play no small part in later developments.

In early 1800, therefore, the third Mysore Survey was approved, if not fully understood, and Lambton immediately began experimenting with instruments and likely triangles. For what was described as 'a trigonometrical survey of the peninsula' it was essential first to establish a working value for the length of a degree of latitude in mid-peninsula. Like those expeditions to Lapland and Ecuador, Lambton would therefore begin in earnest by planning a short arc in the vicinity of Madras. It was not, though, until April 1802 that he began to lay out the first base-line which would also serve as the sheet-anchor of the Great Trigonometrical Survey of India.

The delay was caused by the difficulty of obtaining suitable instruments. Fortuitously a steel measuring chain of the most superior manufacture had been found in Calcutta. Along with a large Zenith Sector (for astronomical observation) and other items, the chain had originally been intended for the Emperor of China. But as was invariably the case, the Macartney Mission of 1793 had received an imperial brush-off and Dr Dinwiddie, who was to have demonstrated to His Celestial Highness the celestial uses of British-made instruments, had found himself obliged to accept the self-same instruments in payment for his services.

Subsequently landing in Calcutta, Dinwiddie had made a handsome living from performing astronomical demonstrations. But he now graciously agreed to sell his props for science, and the chain in particular would serve Lambton well. Comprised of forty bars of blistered steel, each two and a half feet long and linked to the next with a finely wrought brass hinge, the whole thing folded up into the compartments of a hefty teak chest for carriage. Thus packed it weighed about a hundredweight. Both chain and chest are still preserved as precious relics in the Dehra Dun offices of the Survey of India.

A suitable theodolite for the crucial measurement of the angles of Lambton's primary triangles was more of a problem. A theodolite is basically a very superior telescope mounted in an elaborate structure so that it pivots both vertically about an upright ring or 'circle', thus enabling its angle of elevation to be read off the circle's calibration, and horizontally round a larger horizontal circle so that angles in a plane can be read in the same way. Plummets, spirit levels and adjustment screws are incorporated for the alignment and levelling of the instrument, and micrometers and microscopes for reading the calibration. Additionally, the whole thing has to be rock stable and its engineering, optics and calibration of the highest precision. In fact there were probably only two or three instruments in the world sufficiently sophisticated and dependable to have served Lambton's purpose. Luckily he had discovered one, almost identical to that used by William Roy, which had just been built by William Cary, a noted English manufacturer. But it had to be shipped from England, a considerable risk in itself for an instrument weighing half a ton and about the size of a small tractor. And unfortunately the ship chosen was unaccountably overdue.

It had still not arrived when Lambton marked out and cleared his Madras base-line. The site chosen was a stretch of level ground between St Thomas's Mount, a prominent upthrust of rock where the 'doubting' apostle was supposed to have once lived in a cave, and another hill seven and a half miles to the south. Situated on the south-east edge of the modern city, the Mount has since been overtaken by development, but the other end of the base-line is still predominantly farmland and scrub as in Lambton's day. Having cleared and levelled the ground and aligned the chosen extremities, Lambton commenced measurement with Dinwiddie's hundred-foot chain.

By now he had received from England a second chain, but this was reserved as a standard against which Dinwiddie's was frequently checked for any stretching from wear or expansion. Expansion and contraction due to temperature change was a major problem. William Roy of the Ordnance Survey, while measuring his first base-line on Hounslow Heath (now largely occupied by Heathrow Airport), had discarded both wooden rods and steel chains before opting for specially made glass tubes. Lambton in India had no such handy alternative; he had to make the best of the chains. When in use, the chain was drawn out to its full hundred feet and then supported and tensioned inside five wooden coffers, each twenty feet long, which slotted cleverly onto tripods fitted with elevating screws for levelling. Each coffer he now equipped with a thermometer which had to be read and recorded at the time of each measurement. By comparison with the other chain, which was kept in a cool vault, a scale of adjustment was worked out for the heat-induced expansion.

But April and May are hot months in Tamil Nadu. The temperature seesawed between 80 and 120 degrees Fahrenheit. Although Lambton says nothing of the inconvenience of working in such heat, he was worried sick by the variations. After endless experiments he came to the conclusion that a one-degree change of temperature made a difference of 0.00742 of an inch in the hundred-foot length of the chain. But were the locally purchased thermometers sufficiently accurate? And might the temperature not have changed in the interval between marking the measurement and reading the thermometer? Lambton was deeply concerned; measurements and readings were to be taken only at dawn or in the early afternoon when the temperature was as near stable as it got; the thermometers were checked and rechecked, both chains measured and remeasured against a standard bar. Nothing gives a better idea of his passion for shaving tolerances to an infinitesimal minimum than this pursuit of a variable amounting to just seven thousandths of an inch.

To complete the full seven and a half miles of the base-line required four hundred individual measurements with the chain. For each of these measurements the coffers and tripods as well as the chain itself had to be moved forward. It was a slow business even after Lambton's men had been drilled to do it by numbers. The whole measurement took fifty-seven days, and that did not include the time needed for the construction of end-markers. These were meant to be permanent and so had to combine the durability of a blockhouse with the hairline precision required for registering in the ground the actual mark over which the theodolite would be aligned for triangulation.

And still the all-important theodolite had not arrived. In fact report now had it that the ship in which it was stowed had been captured by the French. This turned out to be true. The ship had been conducted into Port Louis in Mauritius and the great theodolite had there been landed and unpacked. Happily the French authorities, when they realised what it was, rose nobly to the occasion. Repacked and unharmed, it was gallantly forwarded to India and arrived in September 'along with a complimentary letter to the government of Madras'.

Lambton could at last begin his triangulation. In late September he took angles from his base-line to pre-selected points to the south and west. The short southern series of triangles down the coast was to determine the length of a degree; it took about a year. Then in October 1804 he turned his back on the coast. Heading west and inland, he would carry his triangles right across the peninsula and then begin the north–south series known as the Great Arc.

Over the next twenty years sightings of Lambton in Madras would be of rare occurrence. As in Canada, he seemed again to have disappeared into a continental void; perhaps after six years on the public stage, he was happy enough to slip back into the wings of obscurity. But the government insisted on progress reports and the scientific world awaited his findings. Lambton's personal papers would disappear with him. Until the young Everest joined him in late 1818 there are few firsthand accounts of his conduct or his establishment. But his reports found their way into the Survey's files and his scholarly monographs into learned journals. Additionally one of his assistants would pen some recollections; and there is the unexpected evidence of two Lambton children, both born while he was working on the Great Arc. As he later admitted, the years spent in India pursuing his obsession would be the happiest of his life.

THREE

Tall Tales from the Hills

When measuring a base-line it was important to discover, as well as its precise length, its height above sea-level. Other heights ascertained in the course of triangulation could then be expressed in terms of this universal standard rather than in terms of individual base-lines. To establish what would in effect be the vertical base of his whole survey Lambton had therefore chosen a site for his base-line which was only three or four miles from the Madras coast and looked, given the lie of the land, to be only a few feet above it. But working out exactly how many was still a matter of some delicacy.

First, on the sands to the south of Madras' famous Marina Beach, the highest tides had been carefully observed and their maximum reach marked with a flagpole. (In 1802 'sea-level' was construed as high water, although later in the century a mean between high tide and low tide would be adopted as the standard and all altitudes adjusted accordingly.) From this flagpole on the beach the horizontal distance to the grandstand of the Madras racecourse, still today hard by St Thomas's Mount, was carefully measured by chain; it came to 19,208 feet. Next, from the railings at the top of the grandstand the angle of depression to the flag on the beach was observed by theodolite. Then the process was reversed with the angle of elevation from the beach to the stand being observed.

The repetition was necessary because Lambton was keen to measure the effect of a phenomenon known as refraction, whereby sight-lines become vertically distorted, or bowed, by the earth's atmosphere. Here was another of those subtle variables which bedevilled geodetic surveying. In particular, refraction would play havoc with long-range observations to distant mountain peaks, although, as George Everest would discover, it also had its advantages.

Having deduced a factor for this refraction, Lambton adjusted his measured angles accordingly. Now, conceiving the sight-line between the flagpole on the beach and the grandstand of the racecourse as the hypotenuse of a rectangular triangle (the right angle being deep beneath the grandstand where a vertical from its railings would intersect with a horizontal from the beach), Lambton had measurements for two of the angles and for one side (the 19,208 feet). Elementary geometry then revealed the length of the other two sides, one of which was the desired elevation of the grandstand above sea-level.

It was important to factor in the height of the flagpole, since its flag, not its base at ground-level, had been observed from the grandstand. Likewise the height of the theodolite's telescope above the ground. And finally, to get the height of the base-line, it was still necessary to deduct the height of the grandstand above it.

This last was done by measuring the stand itself and then 'levelling down' towards the base-line, a comparatively simple process in which the incline was broken into 'steps' whose fall was measured by calibrated staves between which horizontal sightings were taken with a telescope equipped with a spirit level. The base-line itself was not perfectly level and had also involved some of this 'stepping'. So had the original estimate for the distance from the flagpole to the grandstand. All having finally been 'conducted with as much correctness as the nature of any mechanical process will admit of ... I may venture,' wrote Lambton, 'to consider it as as perfect a thing of the kind as has yet been executed.' He then proudly announced that 'we have 15.753 feet for the perpendicular height of the south extremity of the [base-]line above the level of the sea.'

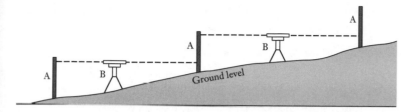

Fig 2   LEVELLING

Over short distances, like the ground measurement for a base-line, the rise or fall of the ground could be measured by horizontal sightings to calibrated staves (A) with a levelling instrument (B) which incorporated a spirit level and a telescope.

Not much attention was paid to this calculation at the time. It had taken several days and much careful planning, but a rise of fifteen feet was no great revelation, and the account of its measurement was buried deep in more technical data about the base-line itself. This in turn was buried deep in a large leather-bound volume whose 1805 publication happened to coincide with news of rather more dramatic elevations elsewhere.

Twelve hundred miles away, beyond the northern borders of British Bengal, a surveyor named Charles Crawford had entered the Kingdom of Nepal in the heart of the Himalayas just as Lambton was laying out his base-line. From around Kathmandu Crawford had got a good look at the Himalayas and, according to an 1805 report of his journey, he had become 'convinced that these mountains are of vast height'.

> ... bearings were taken of every remarkable peak of the snowy range, which could be seen from more than one station; and consequently the distances of those peaks from the places of observation were ... determined by the intersection of the bearings and by calculation. Colonel Crawford also took altitudes from which the height of the mountains might be computed

and which gave, after due allowance for refraction, the elevation of conspicuous peaks.

This sounded most promising. It looked as if Crawford had made the first serious attempt at measuring the Himalayas. Sadly expectations, raised to the snowline in one paragraph, were promptly dashed to the plains in the next.

But the drawings and journal of this survey have been unfortunately lost.

The loss might have been recouped by another writer who happened to have cited Crawford's original findings, but he had done so only in a tantalising telegraphese: 'Double altitudes observed by sextant – allowances for refraction – bearing – computed distance – height by trigonometry – additional height for curvature of the earth – Result, 11,000–20,000 feet above stations of observation.'

The method of operation remained unclear. How, for instance, had the distance of the peaks from Crawford's points of observation been 'computed'? Clearly not in the manner of Lambton constructing his triangle between the beach and the grandstand; but if by horizontal triangulation, this required a base of precisely known length between two points of observation at least twenty miles apart. Crawford's base was rumoured to have been less than a quarter of a mile, and of doubtful accuracy.

Moreover, 'heights above stations of observation' were useless without knowing how high such stations of observation were above sea-level. This information was not given, and an inferred height of about 4,500 feet was mere conjecture. Sea-level deep in the mountains would remain conjectural for the next fifty years, another of the many imponderables which dogged Himalayan observations.

Nevertheless the report put paid to one common misconception. The Himalayas were not a line of active volcanoes. The plumes of smoke which appeared to stream from their

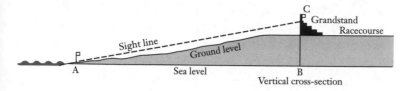

Fig 3

Accessible elevations at short range, like the height above sea-level of the Madras grandstand as measured by William Lambton, could be determined with great accuracy. The distance AB having been established by basic triangulation (see p.8), measurement of the angle of elevation at A (and/or the angle of depression at C) enabled the height (CB) to be calculated to within a thousandth of an inch.

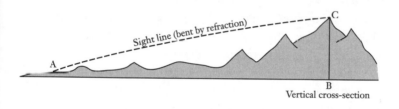

Fig 4

Inaccessible elevations at long range, like the Himalayan peaks as seen from a hundred miles away in the plains, were much trickier to measure. The height of A above sea-level had to have been pre-determined, and an uncertain allowance made for refraction (the bending of sight lines by the earth's atmosphere). Additionally the distance AB was often the result of guesswork or of very doubtful triangulation from too short a base-line whose length was not known with sufficient precision.

summits were simply windblown snow. Additionally, Crawford's attempted measurements represented an important advance on the guesswork which had preceded them. During the next two decades, while Lambton laboured at the triangles of his Great Arc far away in the tropical south, Crawford's Himalayan claims would trigger a wave of both curiosity and controversy in respect of the snowy mountains which, swagged below the Tibetan plateau, defiantly described a great arc of their own along India's northern frontier.

The existence of the Himalayas had been known to the ancients. Ptolemy, the first-century astronomer and geographer, had called them the 'Imaus' and 'Emodi', both words presumably derived from the Sanskrit (H)*ima-alaya*, or 'Abode of Snow'. He showed them as a continuation of the Caucasus mountains running east from the Caspian Sea. Subsequent travellers, like Marco Polo in the thirteenth century, usually trod some version of the ancient Silk Route which, though skirting the north of the western Himalayas, left Tibet and the central Himalayas well to the south. But Tibet had been regularly penetrated in the seventeenth century by Jesuit missionaries from India, and the first convincing account of the mountains comes from one of their eighteenth-century successors. This was the Italian Ippolito Desideri who in 1715 departed Kashmir for Lhasa and was horrified to find, even in late May, the snow deep on the trail and the mountains 'the very picture of desolation, horror and death itself'. 'They are piled one on top of another,' he wrote, 'and so close as scarcely to leave room for the torrents which course from their heights and crash with such deafening noise against the rocks as to appal the stoutest traveller.'

Fifty years later the eruption of British arms into Bengal which presaged the beginnings of the Raj brought more sober appraisals. In the 1760s Lord Clive had commissioned Major James Rennel to survey the territories which, as Colonel Robert Clive, he had so unexpectedly seized. Rennel, the father of the Bengal Survey and its first Surveyor-General, travelled north to the frontier with Bhutan and thence noted several peaks which were snow-covered throughout the year. One in particular stood out; it may have been Chomo Lhari. Although he made no attempt to measure it and considered the hills as outside his field of operations, Rennel did alert the world to the possibility that the Himalayas were 'among the highest mountains of the old hemisphere'.

Curiously, their main rival as Eurasia's highest summit was thought to be not Turkey's Mount Ararat (16,946 feet) nor France's Mont Blanc (15,781 feet), but 'the peak of Tenerife' (12,195 feet). While other quite prominent heights remained uncertain, mainly because they lay so far from the sea and could not therefore be assessed against sea-level, that on the island in the Canaries conveniently rose straight from the Atlantic and lay on a busy sea-route round Africa. Mariners usually possessed sextants, and so the Tenerife peak had been much observed. But at the then accepted height of 15,000 feet, it was still overvalued by almost a quarter. Such was the difficulty of measuring even convenient altitudes.

Rennel had made comparison only with 'the highest mountains of the old hemisphere'. The new hemisphere, or New World, was a different matter altogether. Already the Andes in particular were known to be exceptionally high. Courtesy of that French expedition to measure a degree of latitude on the equator, the peak of Chimborazo in Ecuador had been correctly measured to within a few feet of its 20,700 above sea-level, and so was reckoned the world's highest. That Bhutan's Chomo Lhari was in fact over three thousand feet higher than Ecuador's Chimborazo would have surprised Rennel.

One of Rennel's most distinguished contemporaries was less reticent and actually knew the name Chomo Lhari, or 'Chumalary'. Sir William Jones, a judge in the Calcutta High Court, was unquestionably the greatest scholar England ever sent to India. Dr Johnson had hailed him as 'the most enlightened of men', Edward Gibbon as 'a genius'. Linguist, poet, historian, philologist and naturalist, Jones founded the Asiatic Society of Bengal, whose publications would include Lambton's occasional reports, and he led the field in almost every branch of Oriental studies. It was thanks to Jones that the height of the Himalayas had been added to the agenda of Orientalist research.

'Just after sun-set on the 5th of October 1784,' writes Jones, 'I had a distinct view from Bhagilpoor [Bhagalpur on the Ganges in Bihar] of Chumalary peak . . . From the most accurate calculations that I could make, the horizontal distance at which it was distinctly visible must be at least 244 British miles.' This extraordinary sighting argued strongly for an immense elevation; but Jones also had the advantage of having corresponded with two men who had actually crossed the mountains. They had been sent on separate trade missions to Tibet and had followed an existing and not especially challenging route through Bhutan. But from their reports of latitudes observed and distances gauged, Jones correctly surmised that the mountain wall was many miles thick as well as high. The highest peaks lay well back from the immediate horizon 'on the second or third ridge'. And despite Rennel's caution, after careful study of these and other reports Jones was prepared to chance his arm. He was in fact the first to declare that there was now 'abundant reason to think that we saw from Bhagilpoor the highest mountains in the world, without excepting the Andes'.

In this, as in his other pronouncements on Indian history and philology, Jones's genius lay in divining a truth which as yet defied proof. Such though was his stature that, while some questioned his judgements, more were inspired by them to seek the missing evidence.

Foremost amongst the latter were two cousins called Cole-brooke. Robert Colebrooke was a soldier who in 1794 succeeded to Rennel's post as Surveyor-General of Bengal, Henry Colebrooke an antiquarian and administrator who would become president of Jones's Bengal Asiatic Society and whose broad scholarship mirrored, albeit dimly, that of its founder.

While Colonel Robert would do most of the travelling and would keep an entertaining journal enlivened with delicate sketches, cousin Henry acted as impresario, presenting the findings of Robert and others to the world and pontificating about them.

It was Henry Colebrooke who first took an interest in the mountains. Posted as Assistant Collector to Purnia in northern Bihar, he found himself about ninety miles closer to the snowy peaks than Jones had been at Bhagalpur. During the early 1790s he began a series of observations to try to establish their heights. Assuming their distance to be about 150 miles, and finding their mean elevation to be 1 degree and 1 minute (1°1') above the horizontal (degrees, like hours, are divided into sixty minutes, each of sixty seconds), Henry Colebrooke deduced a height of 26,000 feet.

The question of what this meant in terms of sea-level was not too critical; Purnia lay in the lower Gangetic plain, which was known to be only one to two hundred feet above the tidal reach of the Bay of Bengal. But while heights deduced from observations taken on the plains might be safer in respect of sea-level, they suffered from being much too distant from the snowy peaks and far too vague as to the exact extent of this distance. Baldly stated, the observer either had a good idea of his own elevation and a poor one of the peaks', like Henry Colebrooke from the plains, or no idea of his own elevation but a relatively good one of the peaks', like Crawford in Nepal.

It seemed a no-win situation, but when Henry Colebrooke was posted away from Purnia, he strongly recommended the matter to the attention of cousin Robert. Robert's opportunity would have to wait twelve years. Meanwhile Crawford and others brought back their exciting but scientifically questionable reports of the Nepal Himalayas.

At the time the Kingdom of Nepal afforded the only access to the highest peaks. Much bigger than today, its territory then extended east to Bhutan and west to the Panjab, thus embracing almost the whole sweep of the mountains. If their secrets were to be explored, it had to be through Nepal. Briefly at the turn of the century this looked feasible as the Court at Kathmandu welcomed a couple of British missions, including that to which Crawford was attached. But in 1804 the Anglo–Nepalese treaty of friendship was cancelled and the border closed. The kingdom retreated back into an isolation which, to the chagrin of generations of surveyors and then climbers, would prevent all but diplomatic access for the next 150 years. If the British were ever to get within easy surveying distance of what Jones had so boldly dubbed 'the world's highest mountains', it would have to be by removing some of these mountains from Nepali sovereignty.

Given the pace of British expansion under Governor-General Richard Wellesley's direction, this did not seem too remote a possibility. No sooner had southern India been 'settled' by wresting Mysore from Tipu Sultan in 1799–1800 than the Governor-General turned his attention to the Marathas, a confederacy of rulers who exercised a loose sovereignty over much of the rest of India. There were three Anglo–Maratha wars, and the most sanguinary and significant of them was the second, waged by Wellesley in 1803–4; indeed, it was to 'lay the foundations of our empire in Asia', as he put it. It also laid the foundations of brother Arthur's reputation as an inspirational commander when he won important battles in west and central India including that at Assaye, which he would always recall as a finer victory than Waterloo. As a result of these conquests, Bombay was at last rewarded with territorial gains in western India equivalent to those won forty years earlier by Calcutta in Bengal and four years earlier by Madras in Mysore.

Since Maratha power at the time reached north to Delhi and the Ganges, the opportunity was also taken to extend the territories of British Bengal upriver from Bihar. Called the 'Ceded [in 1801] and Conquered [in 1803] Provinces of the North West', a large tranche of what later became the United Provinces and is today Uttar Pradesh was added to British India. It included Agra and Delhi itself, plus the banks of the Ganges and Jumna right up to where these rivers debouched from the mountains in what was then still western Nepal.

In these newly acquired districts lay Robert Colebrooke's chance to take up the challenge suggested by his cousin Henry. As Surveyor-General for Bengal it was imperative that he map the new territories; and in doing so, he hoped for the first time to push up to the Himalayan foothills in the west and perhaps penetrate them to locate the sources of the Ganges and Jumna rivers.

His resultant survey of 1807–8 had no pretensions to the accuracy of Mackenzie's in Mysore, let alone to the 'correct mathematical principles' in which Lambton took such pride. Colebrooke travelled as much as possible by river-boat. Distances were measured along the bank with a wheeled apparatus known as a perambulator, and bearings were taken to plot locations and occasionally establish latitude, but not with a view to triangulating the territory. It was, in fact, what was called a 'route survey', and its purpose was largely strategic and military. Roads and rivers by which troops could be moved were of the essence; so were fortified towns and other obstructions. The hills were of interest less for their heights than their hollows through which an enemy might invade or, more realistically, a British force advance.

But Robert Colebrooke was well aware of Lambton's work and, while complaining that nothing had been heard of the elusive Yorkshireman for a long time, chanced to mention that it was 'a pity that a survey conducted on such scientific principles is not extended all over India'. Others would soon be thinking along the same lines. Lambton was setting new standards of accuracy which rendered all prior surveys approximate if not redundant. There was no point in wasting weeks plotting triangles with pocket-size theodolites if the Great Trigonometrical Survey with its half-ton instruments and its page-long equations might one day appear over the horizon.

A family man and a happy one, Robert Colebrooke took along on his survey his wife Charlotte, or 'my young lady' as he always calls her, plus the two eldest of the nine children which she had borne him in as many years of marriage. Travelling light was not, therefore, an option. According to Colebrooke's diary, when they forsook their boats his 'equipage consisted of 4 elephants which carried two marquees and 6 private tents; five camels for my baggage; a palanqueen, a mahana and a dooly [different kinds of litters] (the latter two carrying my two children and their nurse), 12 bhangies [bearers for carrying the litters], 12 coolies [all-purpose porters], 12 lascars for pitching the tents, and an escort of 50 sepoys [Indian soldiers]'. Not surprisingly the Colebrookes stuck mainly to their boats.

Throughout his diary Colebrooke happily mixes domestic details with professional notes and extempore sketches. Tigers kept them awake at night, the boats got repeatedly stuck on mud banks, and whole districts turned out to stare at them. Clearly the people had not previously come across a European – let alone a breeding pair complete with offspring. Colebrooke bore it all with grace and humour. Out with his gun of a dewy morning, nostrils flared to enjoy the post-monsoon freshness, the forty-four-year-old Colonel was loving every minute of it. This was the life. It was snipe for breakfast, it was tea with the Nawab, it was India in all its pre-colonial innocence. There was no better place, no better job.

During 1807 the Colebrookes pushed up the Gogra and the Rapti, tributaries of the Ganges, and came within sight of the mountains. At Gorakhpur Robert took his first series of observations of the snowy peaks. Christmas was spent with the small European community in the city of Lucknow. Then, leaving his family behind for what would be a long overland slog, in early 1808 he pressed on to the north-west.

Working along the foot of the mountains, he now encountered thick swamp-forest and the most tiger-infested jungles in India. This was the infamous *terai*, a low-lying belt of tall grasses and towering trees which skirted Nepali territory and was a more effective frontier than the hills themselves. It would claim the lives of a legion of surveyors. Colebrooke himself went down with fever. He took another series of observations to the snow-capped peaks from a place called Pilibhit (near Bareilly at the south-west corner of today's Nepal frontier), but gave up the idea of pursuing the Ganges and the Jumna to their sources. Instead he deputed his assistant, Lieutenant William Webb, to make the attempt.

By April Colebrooke was too weak for anything but river travel. The fever was diagnosed as malaria, complicated by dysentery. He continued to write his journal but the sketches became fewer and the entries shorter. Drifting downriver to Cawnpore (Kanpur) in mid-August he was 'much worse'; and the heat was greater than anything he could remember. Debilitated and delirious, he became obsessed by the monsoon thunder-clouds which piled ever higher and heavier above the river. White-flecked, they towered above him with Himalayan menace. On 12 September he wrote again of an approaching storm. The lightning and thunder continued throughout the night.

> 13th. The weather was so bad as to oblige us to lay all day at Jungeera. Rainy and stormy night.
> 14th.—

With a date and a dash the diary ends. Robert Colebrooke died in the early hours of the morning of the twenty-first, 'a victim to his exertions in the cause of science' as one of his colleagues kindly put it. He was forty-five, not a great age but about the average for Europeans in India at the turn of the century. Life, however delicious, was short. By chance he breathed his last at Bhagalpur, the place whence Sir William Jones had first hailed 'the highest mountains in the world'.

Cousin Henry now owed it to Robert's memory, to the nine fatherless children and the thirty-three-year-old widow, as well as to his own convictions, to present an overwhelming case for the Himalayas. Marshalling the testimony of those earlier travellers into Tibet, of Crawford in Nepal, of Jones, and particularly of cousin Robert and his assistant Webb, he laboured intermittently over his great paper *On the Height of the Himalaya Mountains* for the next seven years.

Like Henry, it appeared from his diaries that Robert too had become convinced that the peaks he had observed from Gorakhpur and Pilibhit were 'without doubt equal, if not superior, in elevation to the Cordilleras of South America [i.e. the Andes]'. At Gorakhpur, Robert had reported that, while a small crowd 'watched me and my instrument in silent astonishment', he had taken angles to two peaks and had deduced for each 'more than five miles in perpendicular height above the level of the plain on which I stood, which must be considerably elevated above the level of the sea'. Five miles was 26,400 feet. He could not be more precise because of uncertainty about the allowance to be made for refraction, that bending of sight-lines by the earth's atmosphere which had so exercised Lambton in Madras. He had used the standard tables showing the deductions to be made, but he had little confidence in them.

Henry, however, was much more confident; for Webb too had submitted observations of a peak which, taken from four different stations on the Nepal border, gave a height of 26,862 feet. Moreover Webb had a name for his peak; he understood that it was called Dhaulagiri, or 'The White Mountain'. So it still is; and within fifty feet, Webb's height is indeed the height now given to Dhaulagiri, the world's seventh highest mountain. But luck as much as science had produced the figure; and Henry Colebrooke then made matters worse by dismissing it as an underestimate; he reckoned Webb's observations gave 'more than 28,000 feet above the level of the sea'. The conclusion of Colebrooke's paper was therefore no surprise.

> I consider the evidence to be now sufficient to authorize an unreserved declaration of the opinion, that the *Himalaya* is the loftiest range of Alpine mountains which has yet been noticed, its most elevated peaks greatly exceeding the highest of the Andes.

First published in the journal of his Bengal Asiatic Society and then widely reported in Europe, Colebrooke's findings caused a minor sensation. But they came up against a certain indifference to all things Indian and were not readily accepted. Armchair scholars, inured to absurd claims from the land of rope-tricks and reincarnation, pooh-poohed these tall tales from the hills. Even recognised authorities with some experience of the Himalayas were not readily convinced. Henry Colebrooke seemed to have overstated his case, to have protested too much. High as they undoubtedly were, the Himalayas were too inaccessible and mountain surveying too approximate to justify his sweeping conclusions.

The most disparaging notice came from the most influential publication. In the *Quarterly Review*, a magisterial journal which until the foundation of London's Royal Geographical Society monitored the course of discovery, an anonymous but highly competent reviewer found Colebrooke's paper 'most curious'. He had no complaint about Colebrooke's methods or his mathematics but, dealing in turn with each of his cited examples, he demolished them one by one. Crawford's Nepal observations were 'of very little value' because his bearings, distances and triangles were unknown. Robert Colebrooke never got nearer than ninety miles to any of his measured peaks, nor did Webb to Dhaulagiri, and nor had Henry Cole-brooke at Purnia. Even assuming that the supposed distances were nevertheless correct, the observed angles of elevation, typically about one to three degrees above the horizontal, were too small for confidence. For every error of one second of one minute of a degree (so 1/360th of a degree) in either the instrument or the observation, fifty feet would be added to or subtracted from the supposed height.

And then there was the problem of refraction. The table of allowances which Robert Colebrooke had used was deduced from astronomical observations. It was never intended for ter-restrial observations at such low angles over such long dis-tances. The *Quarterly Review*'s contributor went into this problem in some detail. Peering out across the English Chan-nel the people of Dover could sometimes see the houses of Calais standing proud of the sea, and at other times, when the atmosphere was equally clear, they could not see them at all. Whale fishers moored off Greenland had noticed the same phenomenon, with snow cliffs appearing and disappearing above the horizon according to the state of the weather and the position of the ice. The whalers called it 'ice-blink' and reckoned that objects thirty miles 'beyond the limit of direct vision' could yet be clearly seen when conditions were favour-able. Temperature, humidity and even the time of the day all seemed to affect the amount of refraction, and in one case it had been found to increase angles of observation by over four degrees. Given that in India the difference of temperature between points of observation in the plains and the ice-encrusted pinnacles protruding above the clouds might be a good 100 degrees Fahrenheit, the properties of refraction could only be guessed at.

In short, Colebrooke's advocacy was fatally flawed. His facts were 'insufficient', his data 'incorrect', his conclusions 'hasty'. 'On every consideration therefore,' intoned the *Review*, 'we conceive that we are borne out in concluding that the height of the Himalaya mountains has not yet been determined with sufficient accuracy to assert their superiority over the Cordilleras of the Andes.'

Colebrooke's response, if any, is not known. But already Webb and a younger generation of surveyors were addressing these criticisms with fierce determination. Their professional competence had been questioned and the honour of their service was at stake. New measurements were being attempted from much more favourable locations, and new proofs would soon be adduced. For by the time that the *Quarterly Review*'s findings had circulated in India, that long-awaited access to a section of the Himalayan glacis had been opened up. Courtesy of the 1814–15 Gurkha, or Anglo–Nepali, War, a slice of the Himalayas between Dehra Dun and the present Nepal frontier had at last been detached from Nepali sovereignty. The highest peaks in the central Himalayas remained shrouded in a haze of Nepali xenophobia, but at their western extremity, in the newly acquired states of Kumaon and Garhwal, there lay lesser giants which, if convincingly measured, could establish the primacy of the Himalayas.

And then there was Lambton. In a sense the exacting standards which he was setting in the extreme south lent weight to criticisms of the rough-and-ready methods and the hit-and-hope calculations adopted by his Himalayan contemporaries in the north. If Lambton could satisfy the demands of European scientists, indeed exceed them, then those same scientists felt entitled to expect equivalent standards of accuracy in respect of claims for the Himalayas.

At a time when the height of Mont Blanc was still unknown to within a thousand feet, this was asking a lot. But there was good news as well as bad. Lambton's Great Arc of the Meridian, having been carried down to the tip of the peninsula, was now heading north. A thousand miles of hill, forest and plain, much of it not British territory and some of it deemed quite impossible to triangulate, still separated the Arc from the

Himalayas. Nor was there any plan to extend it that far. But already Lambton was demonstrating that in the delicious precision of his triangulated distances and heights lay the key to measuring mountains.

FOUR

Droog *Dependent*

Crossing southern India by rail one passes isolated hills which, seemingly dumped at random, are often composed of colossal boulders. They look geologically misplaced, like trophies gathered from afar by some forgotten race of megalithic hoarders. To the upland peninsula's otherwise monotonous succession of wide fields and parched pastures they lend an outlandishness which can be disconcerting. Roused by a vague sense of unease, you look around for giants.

In Karnataka such hills are known as *droogs*, and many once featured as military redoubts during the Anglo–Mysore wars. Similar fangs of rock in the low-lying plains of Tamil Nadu poke through the lush carpet of paddy fields west and south of Madras. The most impressive, like those at Jinji and Trichy (Tiruchirapalli), host impregnable forts and are crowned with tiny windswept temples.

Bangalore-bound from Madras on the Brindavan Express, it occurred to me that a surveyor, invited to design the perfect terrain for triangulation, might well have come up with a landscape model very like the countryside of northern Tamil Nadu or Karnataka. Large, fairly level plains dotted with *droogs* at convenient distances were just what the triangulator ordered. Enjoying a fine, clear climate, not over-endowed with forest, not too densely populated, and yet affording ample supplies, this slice of peninsular India was the ideal place to field-test a trigonometrical survey.

After twenty years' experience in less favoured districts, William Lambton would probably have agreed. But he had not seen it that way in 1803. As he began constructing his first triangles west from Madras to Bangalore and on across the width of the peninsula, he anticipated only difficulties. *Droogs*, for instance, were not necessarily where he wanted them, and when they were, they were not always available. Barely a hundred miles inland from Madras he was obliged to realign the whole northern edge of his chain of triangles. A party sent on ahead to erect a flag at a place called Narnicul had found the desired *droog* defended by 'men with matchlocks, swords and daggers'. They ridiculed the written instructions of the nearest British official and insisted that they held the place in the name of their local 'poligar' or baronial chief. Clearly in India as elsewhere the planting of flags had territorial connotations. Lambton's willingness to adopt whatever flag was locally acceptable made little difference. Eventually the Survey would opt for other sighting marks of a less contentious nature, like a sturdy sapling or a basket atop a pole.

Lambton's men beat a speedy retreat from Narnicul only to be denied again at the next *droog*. This time the reason given was 'that, as it commanded a view of [the poligar's] habitation, his women might be exposed to view'. Such accusations of voyeurism would be another recurrent problem. Elevations invariably commanded the privacy of someone's home, and it was soon common knowledge that the survey's instruments had the ability to magnify distant objects, persons or parts of persons to very intimate effect. Worse still, these monstrous machines not only magnified the object but inverted it. Respectable wives and daughters going about their domestic duties were being upended by perfect strangers with lascivious intent; cherished temples were being casually overturned; and if a well ran dry, it must be because the same troublemakers had tipped it upside down. Nor was it much good these injured people directing Lambton's men to some less objectionable vantage point well out of harm's and harems' way. 'I must place myself on such hills as will descry preceding and succeeding points,' expostulated one of his assistants, 'and these in a hilly tract like this are generally the highest and almost everywhere the stronghold of a poligar.'

The local people were unimpressed; and as for explanations about making maps or measuring the earth, what did these strangers take them for? Maps were made by pacing the roads with pen and paper, not by sitting on hills and tinkering with machines. Besides, everyone knew that if you needed to travel somewhere you found a man who knew the way. Why, even the surveyors were always asking for directions.

Coincidentally, a member of the Madras government's finance committee was reported to have made exactly the same point. 'If any traveller wishes to proceed to Seringapatam [Srirangapatnam], he need only say so to his head palanquin bearer, and he vouched that he would find his way to that place without having recourse to Lambton's map.' The speaker was objecting to the heavy expenditure involved in a trigonometrical survey. Luckily others, including the Governor of Madras, considered the enterprise 'a great national undertaking'. Admittedly the Governor was one of those who were slightly mystified by its geodetic significance; but scientific opinion was evidently impressed and, if the reputation for enlightened government of British India, and Madras in particular, might thereby be advanced, so be it.

Lambton was nevertheless expected to operate on what Robert Colebrooke, with his flotilla of boats and his circus of marquees, elephants and camels, would have considered a wretched shoestring. 'Tents:' ran Lambton's list of sanctioned equipment, '1 Marquee, 2 Private, 1 Necessary, 1 Observatory.' The 'Marquee' was both his office and living quarters; the two 'Private' were needed for storing baggage; the 'Necessary' was for his commode – very necessary given the incidence of dysentery; and the 'Observatory' was for his Great

Theodolite. There were no tents at all for his men, who presumably sheltered beneath whatever they could rig up by way
of a canopy.

Precisely how many men were attached to the Survey in its
early days is uncertain. For carriage purposes Lambton was
initially allowed bullock carts and porters sufficient for the
tents and instruments plus two messengers, eight lascars, two
water-carriers, a carpenter, a blacksmith and an interpreter.
This complement, perhaps forty in total, would soon double.
It was found, for instance, that moving a delicate instrument
like the Great Theodolite by bullock cart was not good for it.
The badly rutted roads caused constant vibration and seldom
went anywhere near the pre-selected heights. Infinitely prefer-
able were porters, who were trained to treat their load with
the care it deserved. But for half a ton of machinery that
meant having at least two relays, each of twelve men, dedicated
solely to this job. A military escort was also found necessary,
partly to overawe hostility as in the case of the poligars, and
partly to prevent the theft of instruments whose brass fittings
were easily mistaken for gold. The escort comprised the Indian
equivalent of a sergeant, two corporals and two dozen privates,
another twenty-seven mouths to feed.

Additionally many of these men – including, it seems,
Lambton himself – might summon their families whenever the
Survey made a prolonged stop, usually to measure a base-line.
If one of Lambton's two European assistants happened also
to be present along with his own train of dependants, the
concourse would be considerable. To men who spent most of
their working lives humping loads up hills in the back of
beyond, such tented gatherings in the open country were a
welcome relief. They were both the social and professional
climax of the surveyor's year. Discipline could be more relaxed;
base-line romances became a cliché of Survey life.

In 1804, having carried his triangles from Madras to
Bangalore, Lambton pushed on to the west, leaving the

measurement of his second base-line to his senior assistant, Lieutenant John Warren. Warren, a fellow-officer in the 33rd Foot, had transferred from Mackenzie's topographical survey much to the latter's regret. His easy-going charm cemented a lasting friendship with Lambton and, as another self-taught astronomer and mathematician, he enjoyed his superior's complete confidence. Nor did he disappoint. The Bangalore base-line took forty-nine days to measure and provided triumphant vindication of Lambton's meticulous methods. For it was found that the measurement of the base by chain along the ground differed from that calculated by the triangulation brought up from the Madras base-line, all of two hundred miles away, by just 3.7 inches in the total length of 7.19 miles. Whatever critics thought about the need for Lambton's Survey, they could hardly be unimpressed by its extraordinary accuracy.

Happily, troublesome poligars were not found west of Bangalore. Across the Karnataka plateau, the going was easier and admitted of the largest triangle yet measured. It is a good fifty miles from Savendroog on the outskirts of Bangalore to Mullapunnaletta, a hill just west of what Lambton calls 'the Great Statue' at Sravana Belgola (a prominent Jain figure, bolt upright and stark naked, which is still the world's largest monolithic sculpture). Yet in the excellent visibility the hilltop statue and then the survey flag were clearly sighted through the theodolite's telescope, and this line duly formed one side of a giant triangle.

Under such conditions a trigonometrical survey could move much faster than its topographical counterpart. Mackenzie, whose relations with Lambton were more correct than warm, was urging forward the men of his Mysore Survey so as to reach the west coast first. The breadth of the peninsula was 'much wanted', as Mackenzie put it, and he was 'very desirous of having this closed first by our Survey for early communication to England'. Specifically 'it would give me great pleasure if [reaching the coast] was effected before Captain Lambton'. In fact, his surveyor who was nearest to the coast was to make a dash for it as soon as Lambton hove into sight; but *'do not,'* Mackenzie underlined, *'mention this to anyone whatever as I confide in yourself alone.'*

Mackenzie's men had been in the field two years longer than Lambton's. It seemed only right that their Mysore Survey should have the honour of crossing Mysore first and so completing the first trans-peninsular triangulation. But pushing their perambulators, pausing every few miles to plot relevant features on their plane-tables, and operating with inferior instruments and much shorter triangles, they were at a serious disadvantage. In long strides like that from Bangalore to Sravana Belgola, Lambton swept past them unnoticed.

There is no evidence that he was responding to Mackenzie's challenge, or was even aware of it. He continued to insist on observing each of his angles on at least four separate occasions. And each time he continued to insist on the theodolite being reversed at least once: from two readings taken on opposite sides of the three-foot circle, or calibrated dial, he could allow for any inaccuracy in the original calibration of its degrees, minutes and seconds (so small that they had to be read with a microscope); such errors could then be rectified by taking the mean of the two sets of observations. Indeed it seems probable that, but for that 'old accident' with the Canadian eclipse, he would also have insisted on separate readings of every angle with first the left eye, then the right.

Likewise he always insisted on taking measurements from all three angles of every triangle. Mathematically, if two angles were known and due allowance made for spherical excess, the third could be calculated. But Lambton strongly remonstrated against any such short-cut. Every angle must be measured. It was the only way to detect errors and it was the only way to discover those variations in the spherical excess so vital to geodesy. Taking the third angle usually meant another long march, the scaling of another scorching *droog*, and then waiting days for perfect visibility – all in pursuit of a value already known to within the smallest fraction of a second of a minute of a degree. But Lambton would not be hurried; and the scientific establishment, if not the government, approved. As the eminent Scottish mathematician and geologist John Playfair would quaintly put it, 'Lambton has no appearance of a person who would save labour at the expense of accuracy.'

To complete the longitudinal arc across the peninsula from coast to coast, it remained to carry Lambton's triangles up to the crest of the Western Ghats and down to the sea. The Ghats, a formidable range which runs the entire length of India's west coast, represented the Survey's first mountain challenge. The dense forests and choked ravines of the Ghats were even more impenetrable than the Kistna-Godavari jungles which would so impress George Everest; and their moist malarial climate, at its worst in the post-monsoon period when Lambton made his final push, was deemed even more lethal. In a note 'Regarding Diseases of the Malabar Woods' a military engineer who had lately been based in the region warned of its extreme 'unhealthfulness' and appended some preventative tips which Lambton might have done well to disregard. They were:

> To wear flannel next the skin and on the feet, more particularly while asleep; to lie high from the ground, and keep a fire in the house or tent during the night; not to walk out while the grass is wet with dew; to smoke tobacco while the air remains damp; and to take regularly as much exercise as the strength will, without feeling fatigue, admit of.

Should the traveller, despite following these injunctions to the letter, contract a fever, the cause was probably 'an enlargement of the spleen'. The only hope then was 'hot bathing, keeping the body bare, and taking continued and even fatiguing exercise'. That and, of course, frequent induction of those panaceas of Anglo-Indian medicine, 'opium, a moderate quantity of the best wine, and a free use of spices'.

Unlike George Everest's reports, Lambton's do not dwell on such dangers. The misfortune of a lost flagman goes unrecorded and the need for extensive tree-felling is merely a 'difficulty'. Tigers are never so much as mentioned; and Lambton himself seems not to have suffered a day's illness in his life. Like 90 per cent of his fellow-countrymen in India he almost certainly caught either malaria or dysentery or both; but he says nothing, perhaps because he saw no need to advertise the fact or perhaps because he feared giving the authorities a pretext for curtailing his labours.

It is known, though, that his men were not immune and that his second assistant was indeed overtaken by the 'Malabar ague'. Nearly dead (from the cure if not the ague), he was invalided home. But India's loss became England's gain; for Henry Kater would go on to become one of the most distinguished physicists of his age, a leading light of the Royal Society, and the inventor of 'Kater's pendulum' and the prismatic compass. In 1823 he would send his old boss news of a miniaturised theodolite which he had just designed; it was as accurate as their 'Great Theodolite' but could fit in a box no bigger than a suitcase and be carried by one man. Sadly Lambton was himself in a box beneath the turf of Hinganghat by the time the letter arrived.

Instead of filling his reports with human detail, Lambton stuck to science. His two stations atop the Western Ghats gave heights above the level of the sea as measured back at Madras of 5,583 and 5,682 feet. From that fifteen-foot slope to the grandstand at the Madras racecourse he had carried his elevations across a subcontinent and up into the clouds. That these were indeed the first precisely measured peaks in India became clear when he continued his triangles down to the coast. Sea-level at Mangalore on the Malabar coast of the Arabian Sea as deduced from that of the Bay of Bengal on the opposite side of the peninsula was found to differ from its actual level by only eight feet. Given the variable and still uncertain properties of refraction, and given the tidal variations, this was as satisfactory a proof of the accuracy of Lambton's vertical angles as the 3.7-inch differential in the Bangalore base had been of his horizontal angles. It was such irrefutable logic which the scientific establishment in Europe would find so painfully wanting in Henry Colebrooke's claims for the height of the Himalayas.

A more surprising discovery was the revelation that the Indian peninsula had shrunk. Against then current maps which, largely based on coastal surveys and astronomical reference, supposed a width of around four hundred miles, Lambton's survey conclusively proved that it was only 360 miles from Madras to Mangalore. A coastal strip of some ten thousand square miles was thus consigned to the Arabian Sea. Thanks to Lambton, British India sustained its greatest ever territorial loss. Fortunately this result mirrored almost exactly that obtained in France when in the seventeenth century a survey had found that Brittany had been represented as protruding over sixty kilometres further into the Atlantic than it actually did. At the time Louis XIV had complained about his surveyors having 'cost me a large part of my territory'. But the precedent thus set exonerated Lambton from similar censure, indeed served to confirm the value of his work.

No base-line was measured on the Malabar coast. The onset of the monsoon in 1805 forced Lambton back to Bangalore and, although he returned to the west coast in 1806, circumstances again prevented the assembly of men, chains, tripods and coffers, let alone the extensive ground clearance, that was essential for base measurement. He regretted the failure but, having completed the first longitudinal arc across the peninsula, he now concentrated on his latitudinal measurement, or great arc of the meridian. (Confusingly, measurements which follow a line of latitude are known as 'longitudinal' because it is degrees of longitude which they traverse and measure – and vice versa.) With the Bangalore base as its starting point, the triangles of what Lambton later called the 'Great Indian Arc' were extended north about a hundred miles to where British territory ran up against that of the independent Nizam of Hyderabad, and then south towards Cape Comorin (Kanya Kumari) at the tip of the subcontinent. The next base-line was in fact measured on the Great Arc near Coimbatore, about 140 miles south of Bangalore, in 1806. In its length of over six miles the difference between the triangulated measurement carried from Bangalore and the actual measurement on the ground came to 7.6 inches.

Another long stride of similar distance would carry the arc down to Cape Comorin in 1809; and another base-line measured near Tirunelveli (south of the ancient capital of Madurai) would produce an equally satisfactory result. But any celebrations over the successful completion of the southward arc were marred by anxiety. For in the meantime, with a particularly sickening crash, disaster had struck. What even the phlegmatic Lambton would concede to have been a catastrophic accident had thrown the whole enterprise into jeopardy.

The trouble had begun amongst the sea of waving palm fronds which blanketed the flat delta region of the Kaveri river east of Tanjore (Thanjavur) in southern Tamil Nadu. In late 1807 Lambton had left the Great Arc to conduct a parallel triangulation down the east coast from Madras. This meant in effect continuing that short arc measured in 1802 to establish the length of a degree as a preliminary to his whole survey. The government was now urging the need for a web of triangles covering the whole peninsula as the basis for a map rather than just the chains of triangles, running north–south and east–west, which promised most to the geodesist. Extending a series down each coastline and then filling in the triangles between these and his Great Arc looked to be the best way of meeting this demand. Hence Lambton's progress south from Madras in 1807–8.

All went well until he ran out of hills and *droogs* in the wide and tree-choked Kaveri delta. Visibility through the coconut palms was impossible without a major felling programme, and the construction of towers was as yet considered an unthinkable expense. Tamil Nadu is, though, famous for its temples. The Tamil temples are in fact the largest in India, and although their numerous shrines and halls may not be especially lofty, they are usually contained within high walls whose gateways, or *gopurams*, support magnificent stacks of sculpture which soar above all else, palm trees included. The solution to Lambton's problem was spectacularly obvious.

There were good precedents for using ecclesiastical buildings. In France surveyors had scaled the towers of several cathedrals including Notre Dame in Paris and, en route to a junction with William Roy's triangles across the English Channel, that of Rouen in Normandy. Roy himself had measured angles from a platform erected around the ball and cross on top of the dome of St Paul's Cathedral in London; and to obtain a flat working area on the spire of Norwich cathedral, he had actually removed its topmost courses of stonework. No such liberties could be taken with temple structures in India. The temples' brahmins had to be carefully and generously handled. Yet the riot of sculpture made climbing easy, and from atop a *gopuram* the level terrain beneath the sea of coconut fronds meant excellent visibility.

Hopping from *gopuram* to *gopuram*, the survey continued south, but in 1808 even the temples ran out. From that of Kumbakonam Lambton had sighted to the west the tower of Tanjore's Rajarajeshwara temple. He now decided to abandon southward progress and to carry his triangles inland to link up with the Great Arc by way of Tanjore.

The Tanjore Rajarajeshwara, or Brihadishwara, is dedicated

to Lord Shiva and was built in the early eleventh century by the great king Rajaraja I, the founder of a south Indian empire whose power reached from Sri Lanka to Malaya and Bengal. Unlike later temples, its *gopurams* are modest, but the main shrine itself is graced with a pyramidal colossus of stonework, 217 feet high. It is in fact the loftiest and, to many minds, the loveliest temple tower in all India. Here was a unique eminence in every way worthy of Lambton's unique instrument. With ropes and pulleys the lascars were soon hauling the Great Theodolite to its summit.

The topmost capstone of a south Indian temple tower is sometimes called an *amlaka*, because of its round and often ribbed resemblance to a myrobolan fruit. That of the Raja-rajeshwara is more dome-shaped and is carved from a single block of granite, over a hundred feet in circumference and estimated to weigh eighty tons. It is thought to have been originally manoeuvred into its exalted position by way of an earthen ramp specially constructed for the purpose and all of four miles long. Compared to this feat, the hoisting of a mere ten hundredweight of machinery looked simple. The pulleys were attached to the *amlaka*; the vast paved courtyard in which the temple stands left ample room for the rope-pullers to manoeuvre; and to keep the theodolite clear of the statuary a guy rope was attached and then probably lashed round one of the pillars of the courtyard's cloister, as to a bollard.

It was this guy rope which either slipped or, according to George Everest's later account, actually snapped. The theodolite, disregarding such niceties as plumb-line deflection, swung smartly towards the vertical and there met the sloping sides of the pyramidal tower, knobbly with sculpture, in a splintering crash. Luckily the instrument was still in its box, or it would have been flattened. As it was, the box took the brunt of the impact and was shattered by the statuary as the protruding tangent screw of the instrument's three-foot circle punctured the packing. The screw sheared and the great circle or dial, so perfectly cast, so minutely calibrated, and so lovingly handled, was left as bent as a bicycle wheel after a head-on collision.

Lambton seems to have accepted full responsibility. 'The high mind of the late Superintendent [i.e. Lambton] could not brook the idea of being reproached for this accident,' recalled Everest, and 'the circumstances of the case were never, I believe, officially brought to the notice of Government.' Although the instrument was valued at £650 (to which sum two zeros may be added for a modern approximation), Lambton ordered a replacement from England at his own expense and then retired along with the mangled original to the military workshops at nearby Trichy.

> Any person but my predecessor [writes Everest] would have given the matter up as absolutely desperate; but Colonel [Captain at the time] Lambton was not a man to be overawed by trifles, or to yield up his point in hopeless despondency without a struggle. He proceeded to Trichy . . . [and] here he shut himself up in a tent, into which no person was allowed to penetrate save the head artificers.

It was the height of the Indian summer, the same summer which, in the far north, found a delirious Robert Colebrooke dreaming of cool Himalayan peaks as he watched the monsoon clouds build above the Ganges while the sacred river bore him downstream to his death. Lambton, like Achilles, never budged from his tent. But far from sulking, he fussed over his beloved instrument through the dog days and sweat-soaked nights like a doctor fighting for the life of his patient. Outside his men waited, their Survey stalled if not permanently halted and their employment in doubt. Everest continues:

> He took the instrument entirely to pieces, and, having cut out on a large flat plank, a circle of the exact size

that he wanted, he gradually, by means of wedges and screws and pullies, drew the limb out so as to fit into the circumference; and thus in the course of six weeks he had brought it back nearly to its original shape. The radii, which had been bent, were restored to the proper shape and length by beating them with small wooden hammers.

Everest was profoundly impressed by this whole saga. Although relying on hearsay, he would tell the story often, and elsewhere says that the repair work took six months, not six weeks. He also suggests that the restored theodolite, though usable, never again inspired quite the same confidence. Lambton appears to have been satisfied with it and was still relying on it when Everest joined him nine years later. But Everest would contend that the main triangulation of the next section of the Great Arc as it edged north into Hyderabad would suffer from the instrument's failings.

Finding fault with men, as with instruments, came easily to George Everest. It would not be out of character for his criticisms to have been a way of trumpeting the higher standards of his own work without inviting the accusation of having personally disparaged his distinguished predecessor. Lambton, on the other hand, had no time for such games. His only recorded criticisms were reserved for those in authority who attempted to curtail his work. Colleagues and subordinates he invariably supported and they revered him without exception.

Moreover it is unthinkable that he, of all people, would have made do with a instrument which he knew to be other than as perfect as humanly possible. If his later work was found to be to a less exacting standard, the explanation lies in his increasing willingness to delegate the actual triangulation to his assistants, Everest amongst them. This was partly dictated by the need to train a successor and partly so that he himself could concentrate on the more arcane calculations and observations vital to geodesy.

Now perhaps in his late forties and greatly encouraged by recent appreciations of his work from Sir Nevil Maskelyne, the Astronomer Royal in London, Lambton was no longer the tongue-tied stranger from the backwoods of New Brunswick. He was still a slave to science, still immersed in mathematical abstractions, and still largely indifferent to the social wheelings and professional dealings of his fellow officers. But ten years in India had given him the confidence of a man who had finally discovered his life's purpose. He now purchased, as was standard practice, the rank of Major and, when not in the field, set up house and home on the coast. He even started a family. But if less aloof, he remained just as elusive. While colleagues would certainly have preferred to live in society amongst the British in Madras, Lambton had chosen to live in sin amongst the French in Pondicherry.

FIVE

The Far-Famed Geodesist

L ambton's reports, as well as lacking the personal detail which might redeem their Spartan syntax, are also extremely light on dates. Numbers and calculations relevant to his work abound, but the simple digits of day, month and year are omitted. If Everest was later confused about whether the repairs to the Great Theodolite had taken six weeks or six months, it may be because Lambton himself had never bothered to record their completion.

We do have a note of how long was spent in measuring each of his base-lines; additionally we know on which nights he conducted astronomical observations, the dates in this case being an essential part of the data. But how many days he spent fixing any of his trig stations is anyone's guess. As with the mystery of his age and birth, Lambton seems to have rejoiced in obscuring the record. Geodetic formulae are known as 'constants'; to a mind obsessed with pinpointing the permanence of place, time's insidious trickle may have been anathema. Or perhaps covering his tracks through the years was a deliberate subterfuge, another dimension of a retiring and elusive persona.

In 1809, after the catastrophe in Tanjore, Lambton's triangulation of the extreme south was overtaken by a smart piece of British aggression against the Raja of Travancore in what is now the state of Kerala. The Survey's presence proved only a minor provocation in what the British historian Sir Penderel Moon calls 'the least justifiable of the many questionable transactions by which British power in India [was] acquired'. During the few weeks which the affair lasted, Lambton, swapping geodesy for gunnery, served as a military engineer. Then, after measuring his base-line at the southernmost extension of the Great Arc, he retired to Pondicherry to work on his calculations and produce the map of peninsular India which would embody them.

Pondicherry then, as now, was an undemanding billet. The tree-lined corniche, the stucco villas, and an air of social and ethnic *fraternité* recalled its golden age in the eighteenth century as the crucible of French ambitions for an eastern empire. Technically it was still the capital of French India. But the British had held it for fifteen years and would continue to do so until after Napoleon's defeat. With a population of only 25,000 it was already a seedy backwater compared to bustling Madras a hundred miles up the coast. Lambton chose it because its cosmopolitan climate appealed to his assistants, few of whom had been born in Britain, and because it was better suited both to his retiring nature and to his greatly changed domestic circumstances.

When and where he had met the mysterious 'Kummerboo' is not known. Her name sounds vaguely Hindu but in Lambton's will she is described as 'a Moor', or Muslim. An officer's loves were seldom transparent and in this Lambton was no exception. According to John Warren, Lambton 'appeared to disadvantage in mixed companies, and particularly in the company of women'. On the other hand Everest, not much of a ladies' man himself (although also capable of surprises), calls Lambton 'a great admirer of the fair sex'. Presumably it was the trussed and trivia-minded British memsahibs whose company showed him to disadvantage, 'not one of them,' according to Richard Wellesley, 'decently good-looking'. Their shyer local sisters, Kummerboo amongst them, must then have been the fair ones – and the not so fair ones – who drew Lambton's admiration.

It is not improbable that there had been other such liaisons during his long stopovers in the field, athwart a base-line perhaps, or atop a *droog*. We know only of those mentioned in his estate. Kummerboo was provided for because in July 1809, with Lambton in attendance, she gave birth in Pondicherry to his first child, another 'William Lambton'. Likewise we know of 'Frances' who ten years later in Hyderabad bore him a daughter and another son. Frances is described as 'a half-caste', perhaps Anglo-Indian or Franco-Indian. Warren hints that Lambton was debating marriage, presumably to this Frances, at the time of his death, a feasible proposition if we discount the Methuselah of tradition.

All three children were acknowledged by him and were baptised. The younger boy seems to have died in infancy, but William junior, after some elementary schooling, accompanied the Survey when in 1815 its headquarters advanced to Hyderabad and, while still only eleven, was put on the payroll as a '3rd Sub-Assistant'. When in 1818 a rather correct Lieutenant Everest was posted to the Survey at its then headquarters in Hyderabad, he might easily have mistaken it for a crèche. As well as young William and the pregnant Frances, Lambton's menage included Joshua de Penning, his senior sub-assistant; Joshua's natural son Joe, a teenager who was also about to become a 3rd Sub-Assistant; and Joshua's wife Marie who, although only twenty-two, was already accompanied by nine of the fourteen little de Pennings she would eventually bear.

Nor was that all. Lambton's original assistants, Warren and Kater, had since moved on, Kater to distinction in London and Warren to head the Government Observatory in Madras. Four army officers had briefly replaced them and had helped complete the triangulation in the south, but in 1811 they were withdrawn as part of a cost-cutting exercise. That left Lambton with just four lowly sub-assistants, all in their twenties and all of whom regarded his headquarters as their home. They were also all locally recruited and all, to jaundiced British eyes, socially disadvantaged because they had been born in India of at least one non-British parent.

Lambton preferred such company. Kater had been of German birth and Warren was French, though born in Italy and with Irish connections. Joshua de Penning, the most senior of these young sub-assistants, had originally come from a Madras orphanage and was perhaps Dutch by birth. Of the other three, William Rossenrode and Joseph Olliver would both, like de Penning, have long and distinguished careers in the Great Trigonometrical Survey. Moreover both had already fathered, or would soon, sons who followed in their footsteps.

Lieutenant Everest would consider himself very superior to all these 'gentlemen', as he called them (the word was meant to emphasise that they were not, like him, officers). Nor would they for their part easily become resigned to Everest's notions of authority. Lambton they worshipped, but for Everest they simply worked; and if in time Everest would come to think of them and their numerous dependants as his family, it was a family which he had inherited.

Not only did Lambton recruit and train these young men, he also demonstrated the utmost confidence in them. As the Survey faced about and began extending its triangles north from Mysore into the territories of the Nizam of Hyderabad, Joshua de Penning was increasingly entrusted with the Great Theodolite and even with the primary triangulation of the Great Arc. Lambton meanwhile took the field mainly to measure base-lines and to conduct the vital astronomical observations.

Star-gazing was almost as important and certainly more demanding than triangulation. The latter simply fixed the Survey's points of observation in relation to one another and to base-lines. But to orientate Lambton's triangles and to establish the position of his trig points on the earth's surface in terms of latitude and longitude, as well as to detect the earth's variable curvature, it was essential that the survey be anchored by astronomical observations. These were usually made by measuring the angles at which planets and stars passed through the zenith as seen from the more important of his trig stations.

The more such observations from any one trig station the better. At the Tirunelveli base-line in the extreme south Lambton had spent twenty-seven consecutive nights closeted in a tent with his zenith sector (an instrument with a telescope and a giant five-foot 'sector' for observing these vertical angles). The result was more than two hundred astronomical observations for latitude at this one location, the mean of which could be taken to give as precise a value as circumstances allowed.

To establish the distance between two points several hundred miles apart, say at the current extremities of the Great Arc, it was important that the same stars be observed at each place, preferably at the same time and with identical instruments. This was asking a lot of the Indian climate since the monsoon in the north of the peninsula does not coincide with that in the south. It also asked a lot of the Survey's resources and would challenge even Everest's genius for organisation. Nevertheless, Lambton was able to make some important corrections to existing maps. The position given to the great city of Hyderabad, whence he now directed the progress of the Great Arc through the territories of the Nizam, he found to be 'out [by] no less than eleven minutes in latitude, and upwards of *thirty* in longitude'.

The Observatory in Madras, where observations and records stretched back many decades and where the dependable Warren now held sway, was the starting point for all peninsular surveys. Its value in relation to the Greenwich meridian (the zero for longitude) and the equator (the zero for latitude) was the Survey's sheet-anchor. Madras was in effect the Indian Greenwich. And when, as happened, some infinitesimal adjust-

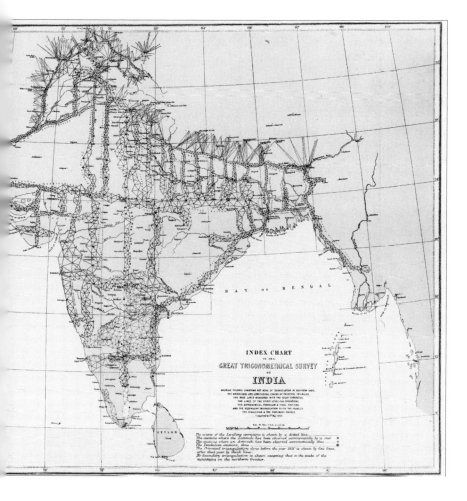

The triangulation of India as completed by 1870 is shown in this index chart of the Great Trigonometrical Survey. In the north George Everest's grid-iron of parallel 'bars' covers the Gangetic valley. Beyond it, the prickly fringe at the extremity of the survey consists of observations to the Himalayan giants.

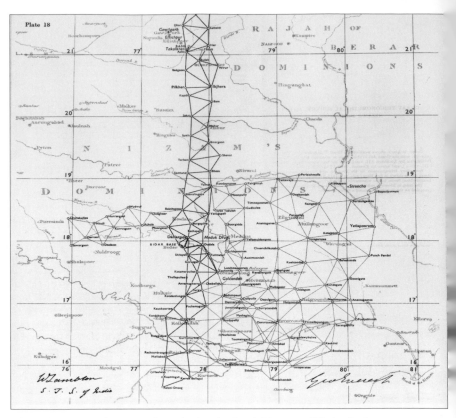

The web of triangulation on the right-hand side of this map of Hyderabad represents George Everest's ill-fated operations in the Kistna-Godavari jungles in 1819–20. Meanwhile William Lambton was pushing north with the Great Arc, i.e. the main chain of triangles up the centre of the map. Both men have signed the map.

Lambton died in 1832 while conducting the Great Arc through central India. The author's rediscovery of his grave in the midst of a squatter colony at Hinghanghat in Maharashtra was a poignant reward for some very amateurish sleuthing.

Before Lambton took the field, surveys (like that of Mackenzie) had limited pretensions to accuracy. Theodolites (ABOVE) were extremely basic, and the survey's equipment (flagpoles, calibrated staves, chain and instrument) could be handled by half a dozen operatives (BELOW).

Lambton's Great Theodolite weighed half a ton and needed twelve men to carry it. Although severely damaged in several catastrophic falls, it was eventually rebuilt and is now housed at the Survey of India in Dehra Dun.

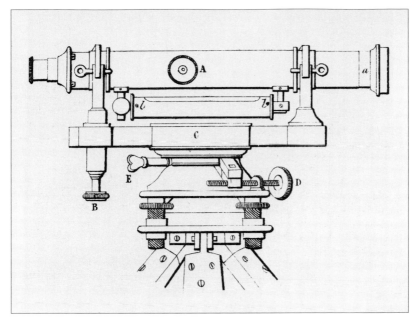

Levelling instruments (ABOVE), incorporating a spirit level and telescope, were used for measuring the rise and fall of the ground along a base-line. Distances could be roughly measured by pushing a perambulator equipped with a mile-ometer (BELOW). Several designs were tried, but the accuracy fell far short of that required for a base-line measurement.

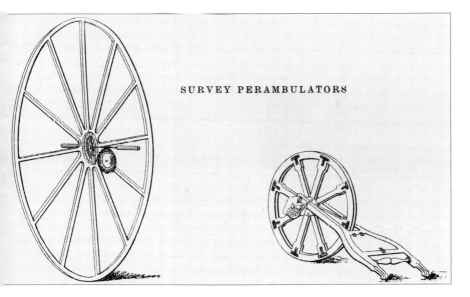

SURVEY PERAMBULATORS

ABOVE The tower of the magnificent eleventh-century temple at Tanjore in Tamil Nadu proved irresistible as a vantage point. But when winching the Great Theodolite to the top, a guy rope slipped and ten hundredweight of precision instrument smashed against the statuary.

LEFT William Lambton, the elusive genius behind the Great Arc, was in his sixties when painted by William Havell in 1822. He had just fathered two more children and won international recognition as a giant among geodesists.

ABOVE A pen drawing of George Everest in 1843 hints at the disciplinarian who so terrorised his subordinates in India. It bears little resemblance to the lionised portraits taken later in life (LEFT).

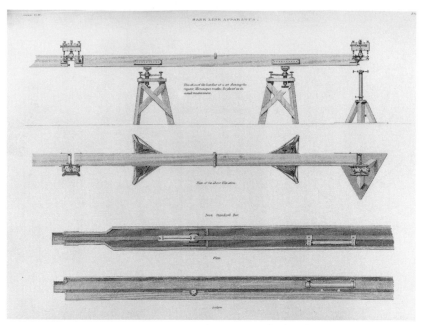

Everest introduced compensation bars for the measurement of base-lines. The drawing shows both elevation and ground-plan of the apparatus with its supporting tripods, plus the microscopes used for registering one bar with the next.

The 1831 measurement of a base-line at Calcutta attracted influential spectators for whose delectation 'an elegant breakfast was laid out in tents'. The awning in James Prinsep's drawing shades the first set of compensation bars. Note in the background the survey tower, one of the first to be built.

LEFT Hathipaon House, near Mussoorie and at the edge of the Himalayas, was where George Everest set up home and headquarters during the final years of the Great Arc. He left in 1843 and the house became derelict soon after; it remains so.

BELOW For astronomical observations George Everest designed this instrument and had it constructed in England in 1830. In 1839 its twenty-four-inch circle was redivided and calibrated in his workshops at Hathipaon, the work being entrusted to Saiyid Mir Mohsin, the Survey's most skilled 'artificer'.

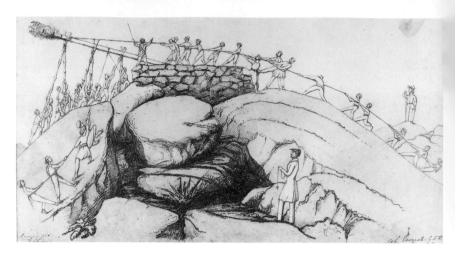

The survey pole so laboriously erected on top of the 12,000-foot Chur mountain would mark the Himalayan terminus of the Great Arc. G.T. Vigne's drawing identifies George Everest as the man in the background (*right*), while one of his assistants, in the foreground, waits with the plumb-line to check the pole's vertical alignment.

Partial access to the high Himalayas became possible after Nepal's 1815 cession of the Garhwal and Kumaon districts. By way of the rope-ladder at Srinagar in Garhwal, surveyors followed the headwaters of the Ganges up to the glaciers and snowy peaks. Oil painting by Thomas Daniell.

RIGHT When in 1820 the Himalayas were at last acknowledged to be loftier than the Andes, the peak of Nanda Devi, known then as 'A2' and measured at 25,479 feet, was thought the highest in the world. It retained this distinction for twenty-five years.

BELOW Forty-foot towers of scaffolding were rigged as observation posts for the preliminary triangulation of the Great Arc across the haze-choked plains of northern India. The theodolite was mounted on a hefty mast (*right*) which stood within, but independent of, the scaffolding with its tented platform.

PERSPECTIVE DRAWING OF THE SCAFFOLDING USED IN THE APPROXIMATE OPERATIONS
OF THE GREAT ARC SERIES IN 1833-4.

DRAWING ILLUSTRATIVE OF THE MAST.

Perspective Drawing

Scale to Mast

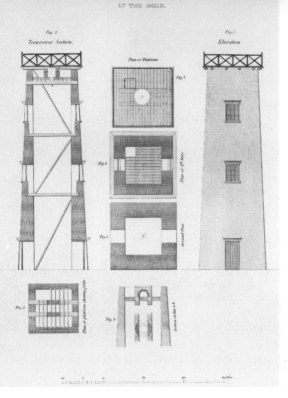

Fig. 2
Transverse Section.

Fig. 1
Elevation

LEFT For the final tri-angulation across the plains, George Everest designed sixty-foot towers built of masonry. His drawings show vertical cross-section, horizontal cross-sections and external elevation.

BELOW LEFT One of the Great Arc's towers, minus its topmost railings, survives at Begarazpur to the north of Delhi. From here the Arc regained the hills as it climbed the Siwaliks towards the Dehra Dun base-line and the Himalayas.

BELOW The Great Theodo-lite, packed in its box, was hoisted to the top of the new survey towers by a specially designed crane. The figures below are manfully demon-strating how to mount and adjust the crane's cross-bar.

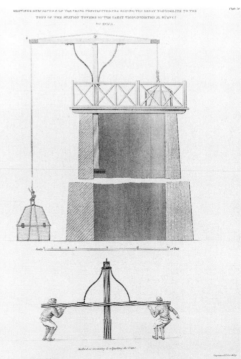

A contemporary view of the 'Musoorie Hills' shows the open valley of the Dun and, beyond it, the Siwalik Hills and the plains. George Everest's home at Hathipaon was on the ridge running off to the right. Engraving by J.B. Allen (1845).

The so-called 'Observatory' on The Ridge at Delhi served as one of the Great Arc's trig stations. Poor visibility here provoked one of Everest's nastier outbursts, although the haze in this near-contemporary photograph is due to fading.

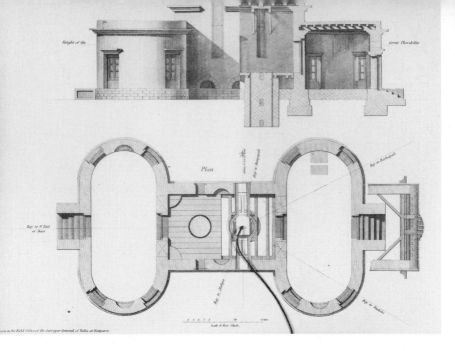

ABOVE Much thought was given to the design of the three observatories (here shown in elevation and ground-plan) from which astronomical observations were conducted at the base-line stations of the Arc. As printed in Everest's *Account*, this illustration has a bit of string trailing from the ground-plan which swivels an in-stitched arrow of card, representing the observatory telescope.

RIGHT Known as 'Strange's Zenith Sector No 1', this instrument of 1866 is similar to (though smaller than) those used by Everest and Waugh to observe for latitude at the baseline observatories of the Great Arc.

Andrew Scott Waugh's 1847 observations of Kangchenjunga from around Darjeeling confirmed it as the world's highest mountain at 27,176 feet. Its reign would be brief. Within a decade it had been demoted to the world's third highest.

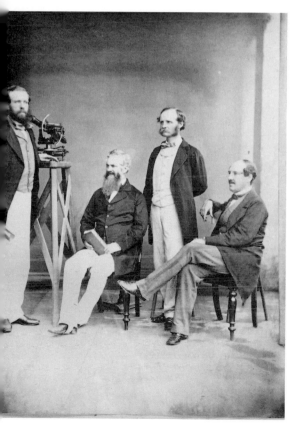

LEFT The tribunal which settled the heights of the Himalayas (*left to right*): T.G. Montgomerie, who first identified 'K2' in the Karakorams, the world's second highest mountain; A.S. Waugh, George Everest's successor as Superintendent of the Great Trigonometrical Survey, who announced the height of what he called 'Mont Everest'; J.T. Walker, a later Superintendent who insisted that the name of Everest was worthy of being 'placed just a little nearer the stars than that of any other'; and H.E.L. Thuillier, the Assistant Superintendent to whom Waugh addressed his judgement on the height of Mount Everest.

An artist's impression of Mount Everest (ABOVE) leaves no doubt as to its pre-eminence. But for the photographer (BELOW) variable weather and a cluster of rival peaks cloud the issue. The pre-eminent summit as seen from Sandakphu is in fact that of Makalu (27,805 feet). Mount Everest, though over 29,000 feet, is the shyer, white peak to its left.

ment was made to the coordinates for the Madras Observatory, Lambton's whole web of triangles had to be realigned. Other such adjustments would necessitate further mind-boggling recalculations. A later writer estimated that the trigonometrical surveying of India involved 9,230 unknowns and produced 'unwieldy equations exceeding anything of the kind ever attempted'. Trial and error, leading to constant refinement, played no small part in the geodesist's science. His situation was like that of a farmer trying to sow his drill evenly with an uncertain number of seeds. When the seeds ran out before he reached the end of the drill, he must needs go back and respace them; and likewise if he came to the end with seeds to spare. Repeating and reviewing past work was as important as prosecuting new work.

For example, the length of a degree of longitude as calculated from that short arc carried south from Madras in 1802 was soon revised when the Great Arc produced a more refined value. That in turn meant that the earliest triangles based on the Madras measurement had also to be revised. As the Arc got longer, other assumptions about the curvature of the earth were reassessed, and these in turn meant more recalculation. Lambton had at first accepted Sir Isaac Newton's figure of 1/230 for the compression of the earth's spheroid at the poles. However this 'constant' proved anything but. It was revised down to 1/304 in 1812 and by Lambton himself to 1/310 when in 1818 his Great Arc had embraced nearly ten degrees of latitude. Everest in turn would come up with his own constants; and every new constant meant recalculating all previous work.

Further complications arose from attempts to refine standards of length. Lambton's anxieties over the elasticity of Dinwiddie's chain increased with every base-line measurement. When the chain unaccountably stopped expanding as measured against the one held in reserve, he became suspicious about the reliability of the latter as a standard. As once before, a hundred-foot brick wall was constructed, tents erected over its entire length, and its surface levelled and polished 'so as to resemble a sheet of glass'. Both chains were then stretched along it, their relative values being assessed by micrometer against a standard bar and then marked against pre-set brass studs, mounted in lead, and set in concrete.

Matters were not assisted when in 1821 a parliamentary committee in England laid down a new standard of length. Lambton had long urged the adoption of such a standard, although he would much have preferred that based on the decimal metre and its derivatives as already calculated by the French. The British ignored this advice. In London Henry Kater reduced the new standard to the scale used by Cary in calibrating the dial of Lambton's Great Theodolite. Lambton had then again to go right through all his measured angles and readjust them.

It was hardly surprising that he had little time for fieldwork. As he patiently explained to the Calcutta authorities, he could not simultaneously continue the surveying and process its results. The calculations involved were so complex that they could be entrusted to no one else. Even producing a fair copy of one of his reports took five months and, because of its highly technical nature, it could not be delegated. Instead he must delegate the fieldwork. If the government were unhappy with this arrangement, they must supply him with a senior assistant; and while they were about it, he also badly needed a doctor and a geologist.

The doctor was required to minister to Lambton's survey parties, who were continually having to quit the field because of fever, while the geologist was needed to sort out his problems with the plumb-, or plummet-, line. Although the Great Arc was providing satisfactory proof of how degrees along the meridian increased in length as he headed north, they were not increasing by any consistent value; in one case they actually decreased. Similar anomalies had been found on the arcs measured in France and England and led cynics to suggest that the surveyors were not as infallible they pretended.

But after checking and rechecking, Lambton was sufficiently confident of his own working practices and calculations to look elsewhere for an explanation. The fault, he supposed, lay in the plummet-line, whose vertical was particularly critical in observations conducted with the zenith sector. It was known that the existence of nearby hills might distort the plummet by attraction, another knotty problem as yet unforeseen by Himalayan surveyors. But Lambton now found that even when he was well clear of hills, irregularities still occurred. The worst example of deflection had been at Bangalore, which he could only ascribe to signs of a subterranean 'vein of dense ore'. And not without a note of triumph he announced that it was this speculation which 'discovered to us an agent unthought of in former days, viz., a disturbing force occasioned by . . . diversity in the density of strata under the [earth's] surface'.

Hence the need for a geologist and hence, in time, a whole new field of geodetic experimentation in which pendulums were used to discover variations in the direction of gravity due to the variable density of the earth's crust. These in turn would reveal that the vertical attraction exercised by mountains was compensated for, and often more than compensated for, by the greater density of the subterranean strata which supported the mountains. As with icebergs, these invisible substrata might extend well beyond that part of a mountain which was visible above ground. Plummet-lines, instead of being attracted towards visible mountain masses, were thus just as likely to be deflected away from them and towards the denser outlying sub-strata, a contradictory and compensatory effect known as 'isostasy'. The Himalayan surveyor was in for more surprises.

In 1818 Lambton learned that he was at last to get both his geologist and his doctor, as combined in the person of Henry

Voysey. In the previous year he had submitted his third report, having in 1815 completed the Great Arc up to Bidar, about eighty miles west of Hyderabad. There he laid out his sixth base-line. The Arc, now of nearly ten degrees (or over seven hundred miles), was much 'the longest that has ever been measured on the surface of this globe'. It had overtaken even that in Europe and, like the Anglo-French arc, 'in grandeur and accuracy [it] must be allowed to exceed anything of the kind recorded in the history of practical science'. No longer merely a curiosity, the Arc had acquired a celebrity and a momentum of its own.

Lambton, too, was becoming something of a legend. In belated recognition of his achievement, the Survey, hitherto variously known as the 'trigonometrical', 'astronomical', or 'mathematical' survey of Mysore – or sometimes simply as 'Lambton's' – was now officially designated as The Great Trigonometrical Survey of India. And in recognition of its having passed beyond the territories controlled from Madras it was transferred from the Madras government to the supreme government in Calcutta and to the personal attentions of the Governor-General. First intended just to cover Mysore, it had since been extended to the whole peninsula, and now in 1818 it was hoped that it might be continued north, east and west at least until lateral chains of triangles could link Bombay and Calcutta.

This meant extending the Great Arc itself still further. Lambton's initial 'foray' into the more unruly territories of the Nizam of Hyderabad had gone smoothly enough. Provided he could evade the 'gangs of plunderers which infest that part of the country when the army is not in the field', he planned to continue following the same 78-degree meridian through the northern districts of Hyderabad and on to Nagpur in central India. 'Should I live to accomplish that,' he wrote, 'there will then be a foundation for extending the survey over the whole of the Deccan . . . through the Maratha dominions . . .

and finally into the upper districts of Hindustan [i.e. north India].' The government endorsed this proposal by suggesting Agra on the edge of the Gangetic plain as a suitable termination.

But Lambton, possibly into his forties when he started on the Survey, had by now been in the field for sixteen years. Whatever his real age, he was beginning to show it. His few remaining hairs were grey, his formidable stamina a wheezing shadow of its old self. Despite his eagerness to carry the work forward, even he was giving serious thought to a successor.

> I sincerely hope, that after I relinquish [the Survey] some one will be found possessing zeal, constitution and attainments wherewith to prosecute it on the principles already followed – It would indeed be gratifying to me if I could but entertain a distant hope, that a work which I began, and which will then be brought to so considerable a magnitude, should at some future date be extended over *British India*.

The hint was taken. In 1818, as well as Dr Voysey, Lambton was also awarded the services of a senior assistant; and it was thus that, on Boxing Day of that year, there rode into Lambton's no doubt riotously festive compound in Hyderabad a clean-shaven and mustard-keen Lieutenant George Everest.

By his own account, Everest approached 'the great man' with deference. Simple manners and reclusive habits did not mean that Lambton was indifferent to recognition and, though long delayed, fame had eventually caught up with him. The Astronomer Royal's letter in 1806 had been followed by an authoritative and highly flattering article of 1813 in the *Edinburgh Review*. Penned by Professor Playfair, it carefully elucidated Lambton's work, applauded his extraordinary dedication, and favourably compared both to those of William Roy, Lambton's original inspiration. In fact Roy and Lambton were jointly hailed as 'doing more for the advancement of general science than had ever been performed by any other body of military men'.

Unaccountably, the scientific establishment in London had not immediately responded. Lambton, unlike Everest, had few distinguished connections and was a stranger to the wiles of self-promotion. But in 1815 his old friend and one-time assistant John Warren had headed home to France. With Napoleon's defeat and the restoration of the French monarchy Warren sought a reunion with his family. Coincidentally he was also reunited in Paris with his old Colonel, Arthur Wellesley, fresh from victory at Waterloo. Reinstated in the French army and recognised as the 24th Comte de Warenne, Warren soon alerted the scientific authorities in Paris to Lambton's work.

Rightly the French regarded geodetic discovery much as the British were coming to regard geographic discovery. They were the pioneers and the arbiters; it was their science, and in Lambton's labours they generously recognised a worthy fellow-worker. In 1817 Jean-Baptiste Delambre, the eminent astronomer, joined Pierre La Place, a leading geodesist whose theorem for spherical excess Lambton had used, in sending him glowing tributes and extending to him the highest honour of a corresponding membership of the French Academy of Sciences.

This recognition finally prodded the Royal Society in London into offering him an honorary fellowship. The East India Company had then followed with what Lambton regarded as a particularly pleasing letter of congratulation from one of their oldest directors. It was especially welcome because the writer was Samuel Davis, once himself a surveyor, who forty years earlier had accompanied a trade mission to Tibet. In fact his were the observations which had alerted Sir William Jones to the great width of the Himalayan chain and which had prompted Jones's still unproven claim about the Himalayas being the world's highest mountains.

To the young Everest, the 'big bald' Lambton seemed a formidable figure. He calls him 'the great man' and claims to have been following his progress for years. A certain eccentricity, which in others Everest would surely have censured, only added to his stature.

I shall never forget the impression which the bearing of this veteran and far-famed geodesist made on my mind when I first saw him . . . at one of our stations; for though we had been in camp together for some days previous, he had displayed no symptom of more than common powers, but seemed a tranquil and exceedingly good-humoured person, very fond of his joke, partial to singing glees and duets, and everything in short which tended to produce harmony and make life pass agreeably; . . . but when he aroused himself for the purpose of adjusting the great theodolite, he seemed like Ulysses shaking off his rags; his native energy appeared to rise superior to all infirmities, his limbs moved with the vigour of full manhood, and his high and ample forehead gave animation and dignity to a countenance beaming with intellect and beauty.

Accolades so fulsome would rarely spill spontaneously from Everest's pen. No doubt his regard for Lambton was sincere, but it was also calculated. In stressing the mystique of the Great Trigonometrical Survey and its founder, he enhanced his own stature as assistant and successor. Nor was anyone meant to infer that Everest's subsequent achievements would owe anything to Lambton's induction. 'He left me,' says Everest, 'in full control of the camp in January 1819 [i.e. within a month of his arrival] to return to Hyderabad; and this was the last occasion of his ever taking part in the work of triangulation.' Lambton had withdrawn because the Great Arc was temporarily suspended until those 'gangs of plunderers' could

be rounded up. But according to Everest it was because he
was too ill to continue.

> These moments of activity [i.e. the joke-telling and the
> duet-singing] were, however, like the last flickerings of
> an expiring lamp. It was evident that he was gradually
> wearing away under the corroding influence of a com-
> plaint of the lungs, attended with a most violent cough,
> which at times used to shake his whole frame as if to
> bursting.

Everest, in short, pretends that he found himself assisting a
dying man. We are thus to understand that he took control
of operations from the moment he joined the Survey. Yet it
would be four years before Lambton did expire, during which
time he was anything but an irrelevance. He would direct his
field parties, calculate and recalculate his computations, visit
Calcutta to lobby on behalf of his staff, and eventually carry
the Great Arc a further 350 miles to the north. With the
enigmatic Frances he also, while debating marriage, cheerfully
started another family.

In 1822, four years after Everest's arrival, Lambton's portrait
was painted by the artist William Havell during a visit to
Hyderabad. Far from showing a wracked consumptive, it
reveals a still genial and almost Pickwickian figure apparently
amused by the artist's attentions. He looks neither seriously
ill nor mightily old.

At the time he sat for his portrait it was Everest, not
Lambton, whose health was shattered and whose career was
threatened. Having been struck down by the Yellapuram fever
during his first disastrous survey of the Kistna-Godavari
jungles in 1819, and then again in 1820, Everest had just
returned from his convalescence at the Cape of Good Hope.
Three years later his condition would become so serious that
he was invalided home. Lambton, on the other hand, had never
yet taken sick leave. Nor, despite that cough, was he showing

any readiness to retire. If anything, Everest's concern may have reinforced his determination to soldier on. To Lambton, as to most Europeans in India, death would come quickly and unexpectedly.

SIX

Everywhere in Chains

George Everest's first catastrophic survey through the jungles of the Kistna-Godavari region had important consequences. For one thing it left most of the Survey's staff chronically debilitated. Even when recovered, they would be highly susceptible to further attacks, usually of malaria or dysentery. Cholera, though less common, was more deadly, sparing neither the fit nor the feeble. And fatalities from other unspecified fevers, although scarcely mentioned in Lambton's reports, become commonplace in Everest's.

Not untypical was the record of another survey party in Hyderabad, a counterpart to Everest's in the Kistna-Godavari jungles, which was conducting a secondary triangulation west of the Great Arc. Under Lieutenant James Garling it had taken the field in 1816 and had made good progress. But 'in 1819 one of Garling's assistants died', notes an official summary, 'and Garling himself died the following year. Conner then came up from Travancore but died within a month of reaching Hyderabad. Robert Young took charge in December 1821, but after two field seasons he also succumbed, and died in July 1823.' Under a man called Crisp the work then 'proceeded steadily', but in 1827 Crisp handed over to Webb and the grim saga began again. 'Webb took sick leave to England in 1829 . . .'

Added to Everest's dismal record of fifteen dead in a single season, such casualty rates cast doubt over the practicality of extending so comprehensive a survey to other regions. The cost of the Hyderabad operations in lives, time and money was deemed excessive. Lambton reckoned that the effort expended in surveying the Nizam's territories would have accounted for an area four times as large, had it been lavished on territories under direct British rule. More worryingly, the government now estimated that the cost of the Great Trigonometrical Survey was running at over £6,000 per year and, with no end in sight, was likely to go on escalating indefinitely. Under the circumstances it was inevitable that the 1818 transfer of the Great Trigonometrical Survey from the supervision of Madras to that of Calcutta occasioned some radical rethinking about both its scope and its priorities.

In this reappraisal the northward extension of the Great Arc was not seriously challenged. Lambton had repeatedly demonstrated the Arc's geographical importance, and international recognition had established its geodetic credentials. The challenge to the Arc in the years ahead would come from the terrain rather than from officialdom. North of Hyderabad, through the heart of central India, lay more hill and jungle, much of it under Indian rather than British rule. There then came the vast Ganges-Jumna plain which stretched north for nearly four hundred miles from Agra to Delhi and on to the Himalayas.

Here trigonometrical surveying looked to be an impossibility. In the early 1820s neither Lambton nor Everest envisaged the Arc ever crossing the plains and reaching the mountains. Agra, where the 78-degree meridian bisected the Jumna and where stately edifices like the Taj Mahal promised commanding views, was regarded as the Arc's likely termination. Thence north there were practically no hills from which to triangulate. Visibility across the plains' interminable patchwork of fields and villages was impeded by a variety of large trees, including the umbrageous banyan, the sacred pipal and the valuable mango, none of which could be casually felled. It was also habitually obscured by a haze compounded of the smoke of several million dung-fuelled cooking fires and the dust kicked up by the world's largest concourse of cattle. The climate promised complications undreamed of in the south, like a cold foggy winter. And the presence of a vast and rather conservative population posed all manner of human problems. Physically the challenge resembled that which Lambton had confronted in the Kaveri delta, but on a much bigger scale, under much trickier conditions, and without the convenience of those soaring south Indian temple towers and gateways.

It looked, then, in the 1820s as if surveys in the northern plains would have to be controlled not by trigonometrical certainties but by astronomical reference. Already those who had succeeded the Colebrooke cousins in their quest for Himalayan heights were experimenting with base-lines whose length was calculated purely by celestial observations at their extremities. Although far from satisfactory, some such method of astronomically ascertained locations was envisaged as the only solution to survey control in the plains.

But if the Great Trigonometrical Survey was to be foiled by the northern plains, there was still plenty of scope for it elsewhere. In addition to pushing the Great Arc forwards to Agra, it was considered essential to 'tie in' Bombay to the west and Calcutta to the east. This was to be achieved by way of lateral or 'longitudinal' triangulations extending outwards along the parallels of latitude. Thus would be established the positions of these cities relative to Madras, and thus would they be linked cartographically as features in the same survey and components of the same map.

Such an all-embracing map, or atlas, was now considered highly desirable. To the British, somewhat in the manner of a tomcat scent-marking its territory, the map would define the area in which they had a personal interest. They called this area 'India', a term then alien to the peoples of south Asia and imprecise even in European usage, and they conceived this 'India' as a distinct Asian entity and hence, by the criteria of colonial expansion, as a legitimate subject of dominion. The map would substantiate this idea by demonstrating their knowledge of the spatial relationships between its component cities, strongholds and geographical features, a knowledge more intimate and accurate than had ever been displayed by the country's inhabitants. And by portraying these relationships in ink on paper, with or without the invisible chains of triangulation, the map would foreshadow their actual linkage by the best chains that the ferrous technology of the age could offer – metalled roads, steel rails and, soon, copper telegraph wires.

More conquests in 1817–19 (the Pindari and Third Maratha wars), which not incidentally cleared a bandit-free path for the Great Arc through central India, were now making British political supremacy a reality throughout the whole subcontinent save for its extremities in Assam and the modern Pakistan. Such independent states as survived within this 'India' no longer posed a threat to British arms, British surveys, or British conceits. In fact the Great Trigonometrical Survey was coming to be regarded as the most explicit expression of the newly won paramountcy.

For reasons of cost as well as of changing ideology, earlier ideas of intensive 'webs' of triangles being spun over the entire territories of, say, Mysore or Hyderabad were being gradually abandoned in favour of an all-India grid composed of crisscrossing 'chains', or 'bars', of triangles centred on the Great Arc. The holes in the grid could be filled in later by cheaper and less rigorous topographical surveys. Lambton himself had been forced to accept this compromise in parts of Hyderabad, and Everest would soon systematise the 'grid-iron' for the whole of India. Whether directly or indirectly ruled, the entire surface of what the British now understood by the word 'India' was, wherever possible, to be speedily subjected to the same standard of measurement.

With scant regard for ancient particularities of environment and culture, a large part of south Asia would thus be engrossed, defined and 'enchained' as one. Critics would rightly see the 'grid-iron' as a symbol of India's incarceration; but to admirers, it symbolised India's incorporation. It was as much about holding peoples together as holding them down; in due course Indian nationalists as well as British imperialists would applaud the work of the Survey.

By its British champions the progress of the Survey thus came to be seen as an enlightened and comparatively bloodless paradigm of the progress of imperial dominion. Its trials became a source of imperial concern, its triumphs of imperial satisfaction. The terminology of the Survey would reflect this. In Everest's reports, each season's operations would constitute a 'campaign', angles would be 'bagged', and mountains, where they occurred, would require 'conquering'.

The lateral, or 'longitudinal', series designed to link the Great Arc to Bombay was the next task entrusted to George Everest when, in late 1822, he returned from his year's convalescence at the Cape. As well as prompting a rethink about the scope and purpose of the whole Survey, the horrors of the Kistna-Godavari jungles had alerted Everest to the need for new methods and practices. Heading west from the Great Arc for Poona (Pune) and Bombay (Mumbai) in October 1822, he began to test out various innovations which would dramatically improve the Survey's prospects.

Compared to his earlier experience, the new assignment was soon proving a joy. 'The face of the country is quite denuded of trees,' he reports, 'here are no jungles to foster fevers, no musquitoes to torment, no banditti to infest the path, no roaring rivers to cut off communications; but a fertile and well-peopled country inhabited by the Mahratta [Maratha] tribes, who are the best natured and kindest of all the natives of India.' Excitements were few and inconsequential. At Achola, a *droog*-like eminence where he established his first station, a pair of striped hyenas, even in those days a comparatively rare species, had established their lair in a cave past which Everest daily strode from his camp to his theodolite. The hyenas refused to move out; it was their territory. Everest refused to alter his route; it was his. The conclusion was foregone. 'Detected lurking in a field of very high corn', one of the 'luckless creatures' was shot.

From Achola Everest moved rapidly west. Speed was important. Lambton had confidently predicted his own arrival in Agra before Everest could reach Bombay. Everest took this as a challenge. He saw himself, as he grandly put it, 'pitted against one whose name had been sounded by fame's trump in every corner of the learned world', and he was determined to forestall him. Less fancifully, he also felt that he had a score to settle.

On his return from South Africa what he calls 'certain trivial circumstances' had embittered his relations with Lambton. He had, perhaps, voiced some of the minor criticisms which he would subsequently put on record concerning Lambton's conduct of the latest base-line measurement; perhaps he had also grumbled about the 'reckless exposure' [to the climate? fever?] for which he held Lambton responsible. More certainly he had taken strong exception to the Colonel's continued preference for his lowly Madras assistants, the 'mestizoes' (as Everest calls them) Joseph Olliver, William Rossenrode and especially Lambton's 'agent' Joshua de Penning. De Penning had been entrusted with carrying northwards, under Lambton's guidance, the primary triangulation of the Great Arc; Everest, on the other hand, a British officer and an English gentleman as well as Lambton's senior assistant, had been fobbed off with the Bombay series. Could he but admit it, it was the unassuming de Penning rather than the fame-trumped Lambton against whom he was pitted; and it was de Penning's arrival in Agra which he must forestall.

Annoyingly the Maharastrian countryside west of Achola permitted no long strides like those by which Lambton had once swept across Mysore. Because of its ridged nature, distant views were blocked and the sides of Everest's triangles rarely

exceeded twenty miles. North of Sholapur, though, the landscape opened out into the flatter, blacker terrain typical of
the Deccan plateau. Broad horizons and fifty-mile triangles
beckoned. The only difficulty was that, for reasons of accuracy,
it was a bad idea to triangulate from a twenty-mile base to a
point three times more distant. Triangles were supposed to
be as near symmetrical as possible. In fact he would later make
it a rule that none which included angles of less than thirty
degrees or more than ninety degrees would be acceptable. If
the size of triangles was to increase, or decrease, it must do
so gradually. To take advantage of the plateau country ahead,
he therefore determined to force the expansion of his triangles
over the last of the ridges.

From his station at a place called Dharoor he sent forward
his flagmen to occupy the most distant point from which
Dharoor was visible. They chose a hill called Chorakullee,
thirty miles away and behind an intervening ridge. Everest
could see nothing over the ridge, but the flagmen insisted that
they could clearly see Dharoor, 'a circumstance which the
wild imagination of my native followers attributed, as usual,
to magic'. Keen, for once, to credit his men's eyesight, Everest
ordered the construction of stone cairns at both stations. If
the sight-line was just brushing the ridge, it might be raised
sufficiently by increasing the height of the two hills.

Stone was piled upon stone; the cairns became towers. At
last, when each was over twenty feet high, a clearer morning
than usual revealed not only the Chorakullee tower but the
whole hill on which it stood. Runners were immediately
despatched to carry the good tidings to Chorakullee and to
order the erection of 'a large mast with a torch at the top of
it'. On the appointed day, at sunset, Everest was perched atop
his tower with the Great Theodolite trained on the horizon.
At first he could see nothing but the intervening ridge. It stood
at seven and a half minutes of a degree below the horizontal.
Then at around 8 p.m. the light at the top of the Chorakullee

mast was seen to break the line of the ridge. 'I watched it rising up the vertical wire [a sighting device bisecting the lens of the theodolite's telescope, and so fine that it was usually made from the thread of a spider's web] till it gradually came to within three minutes of zero.'

The towers, in effect, were superfluous; the flagmen had been right all along; and Everest was now, as he put it, 'fully assured that nature would help me more by the increased terrestrial refraction of the night than any tower less than two hundred feet could do'. Here was a revelatory instance of how refraction, that bending of sight-lines by the earth's atmosphere which Lambton had tried to quantify in respect of the grandstand at the Madras racecourse, fluctuated during the course of the day. As the Himalayan surveyors were discovering, this introduced yet another variable into the vertical triangulation of altitudes; no universal adjustment for refraction could take account of such hourly variation. But as Everest now swiftly appreciated, the same phenomenon could be decidedly advantageous to one primarily concerned with horizontal angles; for points on the earth's surface not apparently intervisible might indeed become so at favourable times of the day and night.

It was the potential of this discovery for nocturnal work which so delighted Everest. While convalescing at the Cape of Good Hope he had investigated, at Lambton's suggestion, an attempted measurement of a short arc of the meridian by the Abbé De La Caille, a French savant and astronomer, in 1751. The Abbé had chosen the Cape because, being south of the equator, it would provide useful corroboration that the southern hemisphere conformed to the oblate shape of the northern hemisphere. In other words it would show that the two halves of the world's 'grapefruit' were identical. Unfortunately it had done no such thing. Indeed the Abbé's calculations seemed to suggest that, although the northern hemisphere might have a flattened pole like a grapefruit, the southern pole must be pointed like an egg. This aberration Everest, following Lambton, rightly ascribed to subterranean interference with the Abbé's plumb-line; but what Everest also noted was that De La Caille had used night-lights for observation. In fact he had found half-burnt timbers within a pile of stones at what he took to have been one of the Frenchman's stations.

Bonfires were far too diffuse and unpredictable as sighting objects, although they were useful for indicating the general location of the sighting object. Flares, with which Lambton had already experimented, were better, but difficult to synchronise because they burnt out too quickly; they were also rather costly. Everest sought a cheap compromise which could be produced locally, and he hit on the extremely simple idea of a terracotta lamp. The 'bulb' was basically a large cup, filled with cotton seeds steeped in oil and ignited. As a shade, a large earthenware urn, thirty inches deep and with a hole in one side through which the light would shine, was inverted over the cup. Any village potter could throw such vessels; they cost next to nothing; and their light could be seen for up to forty miles. They thus 'answered exceedingly well in all but windy weather'.

'I am particular in mentioning this circumstance,' declared a jubilant Everest in his later account of the experiment, 'because it is one which has changed the whole face of the Indian operations.' As well as refraction being at its most helpful during the pre-dawn hours, night-lights were found to be impervious to the haze which proved so troublesome during the day. 'For distances of forty and forty-five miles we can carve a passage right through it, even though it be so thick that the sun appears to set in a sea of molten lead.' True, the surveyor would have to alter his working routine, but as well as changing the clock, he might also change the calendar. Operations need no longer be restricted to those sodden, fever-ridden months during and after the rains when daytime visibility was at its clearest. The cool dry season of November to February, and even the hot dry season of February to June, were suitable, indeed ideal, for night work.

In fact, since Everest found that spying his lamps was greatly assisted by lighting large bonfires either side of them, and since dry firewood for such bonfires was unobtainable during the monsoon, it stood to reason that night surveying could be conducted only during the dry season. Henceforth the monsoon, instead of being the surveyor's open season, would be his closed season, and the risks of blundering into another fever-haunted hell-hole, like that at Yellapuram, would be eliminated.

It was all so gloriously simple. Even the smoke-and-dust haze of the plains north of Agra might yet be penetrated. As Christmas 1822 came and went, Everest continued west towards Bombay. Lambton and de Penning must by now have ended their season and perhaps be back in Hyderabad amongst their motley tribe of children. There could be no question of Everest's not reaching Bombay before they sighted Agra. He was more than halfway there already. The thirty-mile sides of the Dharoor-Chorakullee triangle nicely gave way to the forty-five-mile sides of the next, from one of whose trig points a flare all of sixty-five miles away was clearly sighted.

> Nothing could be more favourable to my progress: all was cheering and *couleur de rose*; and I was busily occupied looking out for my blue lights on this distant station, when a letter reached me from Sir Charles Metcalfe [the British Resident in Hyderabad], communicating the death of my venerable predecessor.

A chastened Everest immediately abandoned his work and began retracing his footsteps. Despite his posthumous diagnosis of Lambton's deterioration, the 'great man's' death seems to have taken him completely by surprise.

It was also unforeseen by others. Dr Voysey, who had made

the Superintendent's health his particular concern, had been sent to Calcutta and on to Agra from where he was now marching south, reconnoitring the ground for the extension of the Great Arc. Meanwhile Lambton and de Penning, forgoing the festive season in Hyderabad, had been in the process of moving the headquarters of the Survey north to Nagpur so as to be better placed for the next stage of the Arc.

They were accompanied by a Dr Morton as stand-in for Voysey. As so often in an age of medical folly, it seems to have been this man's concern for his charge, as much as Lambton's tubercular cough, which finally undermined a legendary constitution. Lambton was used to eating well and drinking copiously, but Morton, after twice bleeding his new charge, prescribed 'the anti-phlogistic system of abstinence from meat and wine'. In their stead he ordained a diet consisting largely of oranges, for which fruit the Nagpur region is justly renowned. Lambton dutifully devoured them in abundance.

But wine was a different matter. Morton had great difficulty keeping him off it and, at the first sign of improvement, Lambton celebrated his recovery by putting himself back on it. On 7 January he downed a pint of madeira and instantly went to sleep. Next day he was far from well, coughed a lot, and spoke little. Morton feared the worst. Hoping to reach Nagpur, the party crept forward and on the nineteenth camped at Hinganghat. Next morning Lambton was late rising. His servant went to rouse him but got no response. 'So tranquilly and calmly had he breathed his last,' wrote Everest, 'that no one was aware of his death.' The far-famed geodesist had died as unobtrusively as he had lived – and as he now lies buried amidst the mud hutments of Hinganghat's squatting millworkers.

'It is now upwards of twenty years since I commenced [the Survey] on this great scale,' Lambton had written shortly before his death. 'These years have been devoted with unremitted zeal to the cause of science, and, if the learned world should be satisfied that I have been successful in promoting its interests, THAT will constitute my greatest reward. In this long period of time, I have scarcely experienced a heavy hour ... A man so engaged, his time passes on insensibly; and if his efforts are successful, his reward is great ... If such should be my lot, I shall close my career with heartfelt satisfaction, and look back with unceasing delight on the years I have passed in India.'

SEVEN

Crossing the Rubicon

L ambton's death in January 1823 meant that Everest, as his only assistant of rank, now took acting charge of the Great Trigonometrical Survey. With it went responsibility for the Great Arc. At last the direction of the world's longest meridional measurement was his; Greenwich would be proud of him. For the next twenty years George Everest would make the Great Arc his personal affair.

At thirty-two, he was younger than Lambton had been when he first launched the Survey, and less obviously a servant to science. No portrait of Everest exists from this period but, in a pen drawing dated 1843, he appears to have retained the Olympian profile of an ambitious youth. Black hair, close-cropped, surmounts the cloudless brow; a frigid stare complements the long cornice of a nose. Glimpsed looming by lamplight over the circle of the Great Theodolite, he may have looked a towering figure. Yet his stature was modest and the imperious brow was belied by a tight mouth and an irrelevant chin. Muttonchop whiskers only emphasised these deficiencies and, in old age, would be allowed to encroach across them, smothering his lower face in a tangle of beard. Then too, lionised by the scientific establishment, he would grow his hair into a mane and thus reward an 1860s photographer with a suitably leonine aspect. As for the lion's roar, he already had it.

In February 1823, having hastened back to base at Hyderabad, he had immediately begun berating those, now at Nagpur, who had been with Lambton at the time of his death. A visit to the graveside in Hinganghat and some active support for the idea of a memorial to the great man would have gone down well with Lambton's mourning companions; but Everest thought only of the instruments and papers which might have been lost to the Survey by the hasty sale of the Colonel's effects.

March brought more soothing news: Everest was officially confirmed as Superintendent of the Great Trigonometrical Survey. The only uncertainty now was over whether there would be anyone for him to superintend. His promotion had prompted a staff crisis as ominous for the prospects of the Great Arc as had once been the Yellapuram fever or was now the haze of Hindustan.

Dr Voysey, his companion in adversity in the Kistna-Godavari jungles, was the first to insist on leaving. Lambton had recently urged Voysey's promotion to Assistant and may have hoped that he, rather than Everest, would succeed him. But the promotion had not been sanctioned and in late 1823 a disenchanted Voysey preferred an uncertain career in England to a subordinate role under Everest. During his recent reconnaissance south from Agra the Doctor had been much troubled by tigers. One, 'a ferocious animal which had carried off five human beings', killed his groom in a lightning attack which Voysey actually witnessed. With Everest, as with tigers, close personal acquaintance evidently argued strongly for a quick retreat. Additionally, Voysey was still suffering from the after-effects of the Yellapuram fever. He would in fact die before reaching Calcutta, let alone England.

William Rossenrode, one of Lambton's part-British sub-assistants, likewise tendered his resignation. Although he was later persuaded to stay on, it was a decision which he would often regret. Fearful alike of fevers, tigers and the demanding Everest, most of the Madras-men who had served Lambton so well for twenty years also sought their release. Everest put this down to the fact that the Great Arc was carrying them ever further from their homes and that, being used to Lambton's indulgent and 'child-like' simplicity, 'they could not immediately transfer their affection to his successor'.

For this misplaced loyalty to Lambton, as for much else, he blamed Joshua de Penning. He had blamed de Penning, still three hundred miles away in Nagpur, for not preventing the sale of Lambton's effects; he blamed him for taking umbrage at the barrage of reprimands which followed; he blamed him for then duly tendering his resignation; and he now blamed him for being so highly regarded by the Survey's Indian staff that they all wanted to walk out with him. De Penning, in short, was a worthless ingrate, a traitor to the Survey and a scientific liability. 'This person,' Everest spluttered, 'had not a particle of mathematical knowledge beyond decimals, the use of Taylor's Logarithms, and the square and cubic root'; indeed it was only by pandering to all Lambton's 'little ways' that he had insinuated himself into 'the absolute mastery of the office and all the arrangements of the Great Trigonometrical Survey'.

Outbursts like this explain the unpopularity of the new Superintendent; during the trying times which lay ahead their language would become even more humiliating and unreasonable. Judged by his reports and correspondence, George Everest may have been the most cantankerous *sahib* ever to have stalked the Indian stage. His pen spat venom; each of his innumerable subordinate clauses was baited with sarcasm and barbed for maximum injury. Yet the effect was mixed, and his sense of outrage was often ludicrously disproportionate to the supposed crime. Like a terrier snapping at the furniture, such attacks could reflect more unfavourably on the biter than the bitten. In time, a few brave souls, noting how the vitriol was neither sustained nor consistent, would come to regard them with a certain affection.

While savaging the inoffensive de Penning in one sentence,

in the next Everest could somehow manage to applaud him as 'highly capable and useful' and express sincere regret over losing him. Far from regarding de Penning's departure as good riddance, he in fact persuaded him to stay on for another year, thereby forestalling the mass resignation of the Madrasi establishment. He then helped him to find an alternative post; and he would eventually coax him back to the Survey's Calcutta office where, accompanied by a now enormous family, de Penning would become Everest's deputy and, belying the classroom maths, his most trusted handler of impossible equations.

More immediately, Everest needed de Penning's help in bringing him up to date with recent progress on the Great Arc. He therefore ordered him and his men to wait out the monsoon in Nagpur, as per the new dry season regime, and then to meet him in October at the Ellichpur (Achalpur) base-line. This was located west of Nagpur within the state of Berar. It had been measured by Lambton in 1822; but the astronomical observations taken by the ailing Colonel required some revision, and the base-line had yet to be connected to the triangulation brought up from Bidar in Hyderabad.

Everest planned to fill in these missing triangles en route to Nagpur. But as ill luck would have it, just before leaving Hyderabad he was caught in a heavy shower and, mysteriously, 'seized suddenly with an uneasy sensation in my loins'. Next day he was running a high fever and aching all over; by the end of a week he was delirious, his limbs paralysed, his skin peeling and his sleep disturbed by 'the most frightful and hideous dreams'. The Yellapuram malaria had returned with a vengeance.

Doctors urged him to head for the sea air at Madras; the jungles of central India, they warned, were a certain death sentence. But Everest was adamant. 'I had made up my mind to resist all these remonstrances, from the fullest conviction that now or never the question was to be decided whether the

Great Arc was to be carried through to Hindustan, or terminate ingloriously in the valley of Berar.'

This was no idle speculation. De Penning and his fretting staff, having already been stuck in Nagpur for nine months, would have taken any further delay as cause to disperse. The mass desertion which had so narrowly been averted would become irresistible and, without the experienced Madrasis whose efforts had so impressed Everest in the Kistna-Godavari jungles, the Survey would be grounded. All momentum on the Great Arc would be lost, and the difficulty of recruiting and training new men might well prompt the authorities to think again about the whole exercise. Everest therefore had no choice. Chattering with fever, in October he headed north for Nagpur and the Ellichpur base-line.

> But it was a desperate resolution; for my limbs being in great measure paralysed, I was under the unpleasant necessity of being lowered into my seat at the zenith sector, and raised out of it again, by two men, during the whole of the operations with that instrument. At the Great Theodolite, in order that I might reach the screw of the vertical circle, it has frequently happened that I have been under the necessity of having my left arm supported by one of my followers; and on some occasions my state of weakness and exhaustion has been such that, without being held up, I could not have stood to the instrument.

For six months he remained a semi-invalid, carted about by palanquin and unable to sleep for more than three hours without being awakened by what he called 'convulsive paroxysms'. Considering that nocturnal observations comprised the main work at Ellichpur, this may not have been a handicap. Everest insists that he made more than three hundred observations for latitude and longitude at each end of the base and then personally visited and took provisional angles at every one of the twenty trig stations north from Ellichpur to Sironj.

This new sector of the Great Arc, over two hundred miles long, sliced through the heart of the subcontinent. The two years which it took were as decisive geographically as they were administratively. But central India's geography is, to say the least, discouraging. It lacks definition and, without the peninsula's familiar coastline or the north's mighty river systems, it defies easy representation. Hills follow no obvious logic and nor does the hydrography. Indeed the tributaries of the Godavari, which drains into the Bay of Bengal, are so entwined with those of the Tapti, which drains into the Arabian Sea, as to make even the east–west watershed a catalogue of contradictions.

North of the Tapti, the deeply scouring Narmada river nearly severs the subcontinent and has often played the role of a Rubicon in Indian history. Delhi's Muslim sultans once ventured across it on blood-and-plunder raids into Maharashtra and the south. Earlier, Hindu dynasties in the peninsula had signalled their ambitions for a pan-Indian dominion by wading its waters to raid and rule up to the gates of Delhi. Following the latter, the sight-lines of the Great Arc were carried across the Narmada valley near a town called Hoshangabad. The British had here established a garrison in whose cantonments the Survey took refuge during the 1824 monsoon.

Voysey and de Penning had by now taken their leave of the Survey. But even they, before departing, had conceded that sighting at night to Everest's terracotta lamps and flares was much easier than scanning the horizon for flagpoles; that it also saved days of waiting for the atmosphere to clear; and that switching operations from the monsoon to the dry season was a masterstroke. Everest himself was greatly encouraged by progress and, feeling better, eagerly planned his next move.

From Hoshangabad the Arc would forge on up the 78-degree meridian, past the city of Bhopal and the great Buddhist stupas of Sanchi (Vidisha), and on to the plain of Sironj beside the Betwa river south of Jhansi. The Betwa flows north-east into the Jumna just before the latter's confluence with the Ganges. It therefore belongs to the same river system as that which drains the Himalayas. Only the great Gangetic basin, with its dearth of hills and its dense mists, would then lie between George Everest and the mountains. Whatever the personal cost, he would soon be able to congratulate himself on having at least wrenched the Arc out of the uncertainties of central India and on to the high road to Agra and Delhi.

Sironj, just two hundred miles south of Agra, had been identified by Dr Voysey in the course of his earlier reconnaissance as an ideal site for the next base-line. Voysey had reported very favourably on the whole line of march. The summits of the hills were mostly bare of vegetation – which must have been a great relief to the 'hatchet-men' – and the camel-eating alligators of the Narmada were either hibernating or extinct. Everest duly reported that he was able to swim his horses and elephants over the river 'with perfect confidence'. From personal experience Voysey had made tigers out to be a much more serious problem. Everest soon agreed; judging by their pug marks, they 'were very large and very ferocious'. To keep them at bay it was sometimes necessary to surround the lampmen throughout their night-long vigils with 'shouts and revelry and the blaze of fires and the discharges of musketry'. Unfortunately the resultant observations, 'which should be made in peace and tranquillity', proved useless.

During thirty years in India Everest himself never actually encountered a tiger. This so impressed his Indian staff that they supposedly credited him with supernatural powers; either that or tigers knew from whom to keep their distance. He was better acquainted with scorpions. At Ranipur, his first vantage point after Hoshangabad, he had a small tent pitched on the cramped summit of the hill while his main tent was erected at the bottom. Having spent the night on the summit, he was next morning presented with a large leaf, stitched so as to make a basket. It was not breakfast but a collection of creepy-crawlies, just a taste of what he had missed; all were scorpions and all of them had been caught by his servants in his lower tent. 'Upon counting them, it appeared that there were, young and old, in number twenty-six – some old gentlewoman, perhaps, with her daughters and her nieces, thus suddenly cut off in their ambitious projects of a suitable settlement.'

Selecting the best hills, but skipping much of the actual triangulation which could be completed at leisure by his subordinates, Everest pressed on for Sironj. In September a recurrence of the fever meant that even in a palanquin he was in constant pain. Again he was tempted to take sick leave; but without measuring the Sironj base-line and declaring the Arc complete as far as the frontiers of Hindustan, he dared not depart.

Through the winter months of 1824–5, the ground at Sironj was cleared, the chains compared, levels taken, and the coffers and tripods set up for the new base-line. It was Everest's first. In a constant state of anxiety, as well as pain, he was trundled from end to end. He insisted on supervising every detail and directing every measurement. Such was the pressure of work that he may well have experienced some kind of mental breakdown. Less charitably, he simply drove himself and his companions to distraction. With Voysey and de Penning gone, the brunt fell on Rossenrode and Olliver, the last of the cadet young sub-assistants recruited by Lambton back in Madras in the early 1800s. Both were about the same age as Everest and both were now highly experienced. William Rossenrode had worked closely with de Penning and Lambton on the Great Arc, while Joseph Olliver had been with Everest on both the Kistna-Godavari survey and the aborted Bombay longitudinal series. Although someone to whom compliments did not come easily, Everest had been unusually generous in his praise of Olliver's work. But in his heightened state of paranoia it was now as if both men were no more than malicious trouble-makers.

The inoffensive Olliver was first in the firing line. He was accused of picketing his horse so close to the Superintendent's tent that it kept him awake all night. This was in direct contra-vention of an order that 'neither men nor cattle should make any sort of disturbance within my hearing.' Did Olliver not appreciate that, in his present state of health, he required every minute of sleep that he could get? Indeed Olliver did, and he denied the charge; but the neighing continued. It went on for three consecutive nights. Someone was obviously enjoying himself. At his wits' end, Everest ordered a guard to pad softly through the camp during the hours of darkness and to turn loose or shoot on sight any stirring quadruped. Peace then reigned, although the mystery remained.

Six weeks later, in November, the unfortunate Olliver was again in trouble. While overseeing some construction work at the far end of the base-line, he had prevaricated over a request that he help Everest with the zenith sector. Everest interpreted this as rank insubordination. His principal sub-assistant, the man whom he had once described as 'my right arm', was summarily arrested, placed under guard, and reported to Cal-cutta. Luckily Everest's superiors knew of his temperament. A reconciliation was urged and Olliver, after apologies, was reinstated.

His case was probably helped by the discovery that it was in fact Rossenrode who had been responsible for the phantom horse which neighed all night. 'You were that person; and it was a horse of yours which created the nuisance,' bellowed Everest when the truth leaked out. Rossenrode at the time was busy working out the mean average of various astronomical values. Resenting the interruption, he responded in kind. Indeed Rossenrode's outburst, being according to Everest 'more befitted to a lewd scold in the purlieus of Billingsgate or Wapping than to a person who had been accustomed to the decencies of life', was deemed more heinous than the original crime. Everest again threatened arrest and disgrace.

It seems, though, that the matter blew over, somewhat literally, when on 10 February 1825 the camp was hit by a minor typhoon. In a foretaste of the winters to be expected in the north, rain and hail accompanied a wind so strong that it flattened all the tents, including that which housed the Great Theodolite. Neither the double guy-ropes nor the ten men who were hanging on to them could do anything about it. As the tent went over, so did the great instrument, crushing beyond redemption 'one of the beautiful Troughton barometers' which happened to be standing nearby. 'Fortunately,' reported Everest, '[the Great Theodolite] has received no other injury than the breaking of one of the lower screws which I have the means of repairing.' But this may have been a misleading assessment of the damage. In another context Everest describes the timber frame of the theodolite as splitting and the whole thing as being 'sadly ricketty'. Almost immediately after the accident the instrument was packed away and no one was allowed to use it. Except for an outing to measure a single angle in 1826, it was not used again until 1835, by which time an expert instrument-maker from England had completely rebuilt it.

In 1825 Everest himself had had no further use for it. With the Sironj base-line completed and his health more precarious than ever, in March he finally indented for sick leave in England. He sailed from Calcutta later that year and, having been assured that his job was safe in his absence, did not reappear in India until five years later.

Allowing one full year for the voyages home and back, five years was still a long absence. To retain his position and at least part of his salary, it was necessary to prove that in Europe, as well as recuperating, he would be busy about the Survey's business. He therefore took with him sufficient documentation to work up the results of his last two years' work and to write an account of it. On arrival in London he also began visiting instrument-makers and indenting for the latest in the way of survey apparatus. Besides theodolites and zenith sectors, he inspected lamps for night work and a sun-reflecting mirror, variously known as a heliostat or heliotrope, for daytime sighting. His employers, the directors of the East India Company, were not immediately impressed by this shopping spree. But Everest had powerful patrons and was able to mobilise the support of the scientific establishment following his election to Fellowships at both the Royal Society and the Royal Astronomical Society. The new instruments were finally sanctioned and he was soon busy driving their makers to distraction with a long list of modifications.

During his absence from India the post of Surveyor-General with responsibility for all Indian surveys other than the Great Trigonometrical Survey fell vacant. Well-placed to press his own credentials, in 1829 Everest secured his appointment to this most senior position and would thus return to India as both Superintendent of the Great Trigonometrical Survey and Surveyor-General in charge of all topographical and revenue surveys. These new responsibilities could have proved a distraction from his triangulations and in particular from the Great Arc. In fact they would stimulate them, since the resources of the other surveys would be increasingly redirected to the Great Trigonometrical Survey.

Home leave proved timely in other ways. At the instigation of Richard Wellesley, now Marquess of Mornington and Lord-Lieutenant in Dublin, the British Ordnance Survey had recently been entrusted with a detailed survey of Ireland. The Master General of Ordnance was Richard's brother Arthur, now Duke of Wellington and Prime Minister. Unsurprisingly, given such patrons, the Irish survey was being conducted on an unprecedented scale, to witness which Everest spent three months in its company. He noted particularly its institutional structure and instruments; and he also observed with interest the thirty officers, the three hundred other ranks and civilians, and the like number of labourers thought necessary for what was indeed an ambitious but basically straightforward trigono-metrical survey of an island rather smaller than the state of Hyderabad.

Clearly, in the light of this Irish revelation, it would not be out of order to request more personnel for his Indian oper-ations and to expect better funding. He also noted how in Ireland the trigonometrical measurement was deemed a pre-requisite for all other surveys, something which in India was often professed but, because of the variety of other demands and because of the existence of other surveys, was rarely prac-tised. Here was the justification he needed, as Surveyor-General, for according priority to the Great Trigonometrical Survey and his beloved Arc and concentrating all resources on them. Additionally he remarked the ease with which the Irish surveyors could get their instruments repaired and checked; an instrument-maker for India was requested and duly supplied. And finally, besides all the other instruments he had ordered, he obtained a double set of 'compensation bars' to replace the chains hitherto used for measuring base-lines.

Compensation bars had been developed by Thomas Colby, the Ordnance Survey's Surveyor-General who was now con-ducting the Irish operations. Because of the familiar problems of chains expanding in the heat and becoming worn with use, Colby had experimented with twin bars, each ten feet long and each composed of a different metal whose different rates of expansion could be made to cancel one another out mechan-ically. Compensation bars thus consisted of two bars, one of brass and the other of iron, aligned side by side, bolted together in the middle, and with a cunning little attachment, like a side-bolted lever, linking the bars at each end. As the iron bar expanded it caused the lever to pivot about its axis on the less expansive brass bar so that a dot marked on the lever's projecting 'handle' remained in the same place. The bars were supported on brass rollers and housed in a wooden case from which the levers with the dots on them protruded. Six such sets of bars comprised the usual complement and, once set up and joined together by their accompanying microscopes, gave a length of sixty-three feet.

Everest was highly delighted with them. Before leaving for India in 1830 he tested the new apparatus in London, finding just the flat surface required at Lord's Cricket Ground. He repeated the experiment in Calcutta soon after his arrival in 1831, and later that year he commenced in earnest the measurement of a six-mile Calcutta base-line.

Calcutta, of course, was nowhere near the Great Arc. But as the seat of government, it was the perfect place in which to impress British India with the demanding nature of the Great Trigonometrical Survey and of the superior science which was being invested in the Great Arc. Additionally, the base-line was a necessary means of verifying the accuracy or otherwise of the work which had been undertaken in Everest's absence.

While he had been on leave, Joseph Olliver, assisted by the rebellious Rossenrode, had been given the task of carrying a triangulation eastwards from Sironj, where the Arc currently terminated. Everest had refused to entertain the idea of those whom he called Lambton's 'mestizoes' being entrusted with the Arc itself, but in this eastward series he had seen a useful way of keeping his men together and of incorporating Calcutta into his grid-iron of triangulation. Known as the Calcutta Longitudinal Series, it was, in effect, the equivalent of the Bombay Longitudinal Series on which he himself had been working when Lambton died in 1823. The distance, however, was longer – nearly seven hundred miles in this case – and the terrain more challenging. Initially it had consisted of formidable jungle, in which one of Rossenrode's sons had died and whence the entire survey had had to be repeatedly rescued in a state of malarial collapse. Thereafter the Survey descended into the Gangetic delta in west Bengal and had progressed to within about seventy miles of Calcutta by the time Everest reappeared. The new base-line would complete this work and, when connected up to it, would reveal its accuracy.

But there were good reasons for supposing that the result would be disappointing. For one thing, Olliver had been denied the use of the thirty-six-inch Great Theodolite and, instead, had had to make do with an eighteen-inch instrument of inferior design. Comparing the telescopes of the two, Olliver found his smaller instrument useless for distances of more than about twenty-five miles. His triangles were consequently cramped and his trig points numerous. Considering that the overall distance of seven hundred miles was much the longest yet undertaken without an intermediary base measurement, Everest can hardly have expected the highest standard of accuracy. Indeed one wonders at his subsequently adopting Olliver's series as the basis for a succession of meridional arcs – the 'bars' of his grid-iron – running up to the Himalayas.

Another problem with which Olliver's men had had to grapple was that of extending a trig survey across a flat landscape covered in trees and blanketed in a soupy haze. Bengal was part of the Gangetic plain, the long-feared obstacle across which trigonometrical surveying was thought impossible. Perhaps unwittingly, Olliver was being used as guinea pig. By sighting at night to lights, Everest had shown that mists could be penetrated, but it was still necessary to attain vantage points from which to see over the leafy canopy. With ladders lashed to branches, one above the other, Rossenrode and his men in the advance party had climbed sixty to eighty feet into the trees. Still they could get no clear view, added to which 'the dread of tigers was so great that the Bengali labourers would at every rustle of the dry leaves throw down the ladders and disperse in all directions'.

Scaffolding towers were tried with greater success, but they required much timber and labour, and it was almost impossible to make them stable enough to observe from. They moved in the slightest wind and, because of the difficulty of managing both lights and instruments on a flimsy little platform ninety feet above the ground, night work was found impossible. Much better were masonry towers, and in this respect Olliver was fortunate. A line of just such towers fringed the northern edge of his series through western Bengal and so provided half of his requirement. The towers had been built early in the century as a part of a primitive, pre-electric telegraph system by which messages, preferably short ones, could be flashed up-country from tower to tower. Everest had himself worked on the telegraph towers before being posted to the Great Trigonometrical Survey in 1818 and, no doubt, had been reminded of them when constructing those outsize cairns on his Bombay series.

The telegraph towers required some modifications for survey use, but the expense was as nothing compared to that of constructing new towers. Nevertheless, after his return Everest found that to complete the Calcutta series it was necessary to build eleven additional towers. Several designs were tried, and the cost was indeed considerable. But crucially it was this experiment which convinced him of the feasibility of continuing the Great Arc north from Agra (as and when it got there), across the plains to Delhi, and on to the Himalayas. Instead of *droogs* and temples and tousled hills, the Great Arc would approach the mountains by way of a chain of specially-built masonry towers. And so, provided the funds were forthcoming, the last great barrier to making the Arc an India-long measurement could be overcome.

Funds depended partly on the East India Company's Court of Directors in London, whom Everest had lately so impressed, and partly on the authorities in Calcutta. Hence the importance of the Calcutta base-line. In the nature of things, most of the Survey's triumphs had hitherto occurred miles from anywhere in places which no one had ever heard of. Locations like Sironj or Ellichpur are rarely to be found even on a modern map. But Calcutta was the metropolis, the 'presidency'. Here at last was a chance to perform before a large and influential audience in the heart of British India.

The preparations were meticulous. Everest had originally chosen an eight-mile stretch of country well outside the city. It ran between two of the telegraph towers, but it was found that three hundred trees would still need to be felled, numerous mud houses removed, and a succession of small ponds filled in. To avoid such extensive site clearance he was persuaded to adopt instead a six-mile line along the Barrackpur road, a straight and level thoroughfare which, now as then, flanks the Hughli river and leads due north out of the city. New towers, seventy-five feet tall, had to be built at either end of the new line and, because the road was a busy one, traffic restrictions were necessary. The left-hand side of the road was closed off; cattle, sheep, pigs, dogs and led horses were prohibited entirely; and, to reduce the dust, carriages were to proceed at no greater than a slow walking pace when within 150 yards of the actual measuring apparatus.

Operations began on 23 November 1831 and were completed two months later after 539 separate settings of the six-piece compensation bars. That averaged out at about twelve settings, or 750 feet, per working day; but occasionally it reached 1,254 feet, 'which is double what was effected on the Irish Survey', according to James Prinsep, the distinguished President of the Asiatic Society. Numerous dignitaries visited the operations which terminated, by way of demonstration, with a remeasurement of the first section. Prinsep was amongst those invited to witness the event.

An elegant breakfast was laid out in tents after the ceremonies of the morning were completed. While

contemplating with admiration the order and precision with which the whole process was conducted,
we took an opportunity of sketching the apparatus as
it stood . . .

Prinsep's sketch shows the bars aligned on tripods under an
awning of tents with one of the towers in the background.
Various items of the apparatus are obligingly scattered about
in the foreground, including a case of compensation bars and,
beside it, the microscope-studded device used to connect up
the bars.

The object of the demonstration was to see by how much
this remeasurement of the base-line's first section would differ
from that conducted two months earlier at the start of operations. Needless to say, the results were most satisfactory. Over
the first seven hundred feet, the remeasurement gave a difference of just 0.026 of an inch, much less than had been found
when trying the same test at Lord's cricket ground. 'It would
amount to about 12 feet between Calcutta and Delhi,'
reckoned Prinsep, 'or 125 feet in the diameter of the great
globe itself.'

Besides measuring the Calcutta base-line and assembling
and testing his new instruments, Everest devoted his first two
years back in India to reorganising the Great Trigonometrical
Survey. The priorities, as proposed by himself and now
endorsed by government, were to complete the Great Arc and
to extend his grid-iron of triangulated 'bars' over the whole
of 'Bengal', a term which had come to embrace all of northern
India from Calcutta to Delhi. This ambitious programme was
to be completed within five years. It would therefore entail
advancing on all fronts at once with logistical support of military complexity.

Under the new instrument-maker from England, workshops
had to be set up in Calcutta and the Great Theodolite brought
in for overhaul. Another highly skilled 'artificer' was attached

to the Survey in the field. Engineers were also essential. Some £77,000 would be needed just for building towers. And then there would be a much higher figure for salaries and expenses. Instead of one or two parties in the field at any time, Everest proposed six: two to work on the Great Arc and four to work on the north–south 'bars' of his grid-iron either side of Olliver's Calcutta series. By late 1832 the work of recruitment and training was already underway, with Olliver and Rossenrode being trailed from hill to hill, like doctors on their rounds, by a bevy of tongue-tied juniors. To process the flood of data and handle the number-crunching, more 'computers' were also essential. Joshua de Penning was recalled from Madras to head a Calcutta office devoted entirely to calculation.

As a result of this reorganisation the Great Trigonometrical Survey would acquire a much more institutional structure, although with a martinet like Everest in command it scarcely lost its personal character. To historians of the Survey the period 1833–43 would become 'the Everest decade'. His age of unquestioned ascendancy and achievement was just beginning. If anything, the Survey's augmentation reinforced his self-esteem and encouraged an even more wilful exercise of authority. During the winter of 1833–4, while he at last headed west to resume operations on the Great Arc, his now considerable field establishment was ordered to head for the hills. The journey took five months. In a flotilla of boats, some of which had to be specially adapted to prevent their sinking under the weight of machinery, the Survey's main establishment of men, instruments and horses sailed up the Ganges, then took to roads and hill tracks to trail up to Mussoorie, seven thousand feet above sea-level on a ridge in the Himalayas.

There Everest had just purchased an estate which he had designated as his new headquarters and from which all operations would be conducted during his remaining years in India. The place, although he had never seen it, never been near Mussoorie, never even seen the hills, was carefully chosen.

From atop the ridge, to the north and to the east, there stretched, in a breathtaking panorama of sublime savagery, the snow-scarred peaks of the Great Himalaya. Everest had finally set his sights on the mountains.

EIGHT

So Far as Our Knowledge Extends

Trying not to be distracted by the snowy spectacle to the north, I once tramped the sward up on the ridge of Mussoorie in search of Everest memorabilia. It was a summer's morning. The grass was still moist and the stunted trees were in leaf. What I took to be Logarithm Lodge turned out to be roofless with thickets of prickly acacia sheltering in the lee of its crumbling walls. Of Bachelors' Hall, another satellite bungalow, there was no sign at all. Perhaps it had been levelled by an earthquake, perhaps it had simply succumbed to the climate. Uniquely these south-facing ranges which buttress Asia's mightiest mountains receive both the considerable force of the Indian monsoon and the icy blasts of a Himalayan winter. Architectural casualties tend to be heavy.

So it was cause for celebration to find that east along the ridge, on a grassy terrace at the edge of the void, the substantial shell of Hathipaon itself was still defying the elements. Frost and thaw had prised the rendering from the stonework; wind and rain had bleached the masonry of its rounded bays and gables. All the glass had long gone, and so had the timber of window frames and doors. Yet here it unmistakably was, the house which Everest called home. At Hathipaon he would devote himself exclusively to the Great Trigonometrical Survey, 'my own darling profession' as he called it; and from this podium in the Himalayas he would conduct the Great Arc to its climax in what he reckoned to be 'as perfect a performance as mankind has yet seen'.

It was hard to associate the house with such triumphalism. Bewhiskered with seedlings, furrowed with cracks and slumped in its hollow on the ridge, it had an air of total self-absorption, like an old man in an armchair gazing out over India. I felt as much a trespasser as the buffalo men. A bearded patriarch with a hefty stick and a boy with a pan of water, they were crouching over a smoky pile of twigs in the middle of the drawing-room floor while their lumbering beasts splashed shit in the hallway and scratched long reptilian necks on the dining-room hatch. What could the buffalo men know of Hathipaon's halcyon days? What could anyone know? The house was still there. The location was unbeatable. One day it might be restored. In 1990 it had been announced that the state government was planning to acquire it and, 'as a memorial of Sir George Everest', to develop it into 'a TOURIST/EXCURSION spot'. Perhaps, if they ever get around to it, they will include a 'Great Arc Experience' with life-size dummies and a Bombay soundtrack. But in its present state of abstraction the house is saying nothing.

Hathipaon means 'Elephant's Foot', although whether this refers to the stumpy profile of one of the flanking hills or to the indentation left between them is unclear. It had been built in 1829–30, when Everest was in England, by a British Colonel who had taken a fancy to this commanding ridge on the edge of the Himalayas. Evidently he foresaw its potential as a bracing retreat from the searing temperatures of the plains six thousand feet below. At the time, along the ridge to the east, the little village of Masuri was already developing into the toffee-nosed township of 'Mussoorie', one of British India's premier hill resorts. The Colonel's investment proved as sound as his house. When he headed back to Britain in 1832 it was reportedly 'at a very heavy cost' that both Hathipaon and the surrounding six-hundred-acre Park estate were purchased by the now Captain George Everest.

The new owner had plans for the place. Within the main

house, the five hundred square feet of deal flooring and decorative plasterwork which had been a drawing room now became a drawing office. Outside he built workshops, a small observatory and extensive storage facilities. Logarithm Lodge and Bachelors' Hall were laid out to house his assistants; other members of his staff would be encouraged to erect their own temporary accommodation within the grounds. The near-perpendicular access track was regraded as a carriage road. Heavy and extremely fragile loads were soon being hauled gingerly up the four-thousand-foot escarpment by cart and porter. The woods sang with saws and the workshops billowed with smoke.

It was hard to imagine such scenes of industry. Edging away from the house, I sat on the grass in the summer silence to admire the view. A goatherd and his flock emerged from the abyss and tripped daintily off to tea with the buffalo men. They were followed out of the void by lazy puffs of cloud which hesitated before being snatched up by the wind, tumbled and shredded as they were bundled across the ridge, and then sent swirling off to the snowfields in the north. Sliding over the grassy saddle or snagged among the conifers, these wraiths of cloud spread a clammy chill which troubled the spirit. As the view suddenly vanished, I fancied that I felt the passage of unclaimed souls, flying the land of reincarnation and hell-bent for Tibet.

Cloud was the mountain surveyor's greatest enemy. It could detain you for days in the most inhospitable places imaginable. Bearing roughly north-west from the ridge, though hidden from Hathipaon itself, stands a handsome peak known as The Chur or Chaur. At 12,000 feet it is not a giant by Himalayan standards. But it stands alone, is easily recognised and is eminently climbable. Surveyors keen to sight the much higher peaks to the north and east invariably adopted The Chur as one of their observation posts.

There Everest himself would spend many an icy hour

waiting impatiently for the cloud to clear. Indeed, if there was one Himalayan peak with which the name of Everest was most commonly associated during his working life, it was this humble eminence. Godfrey Thomas Vigne, a freelance traveller, artist and sportsman, would find him there in October 1834. Everest's tent was perched as near the top as possible and, despite a stove and an ample supply of claret, 'our chief object was to keep ourselves warm,' reports Vigne. He stayed the night and next morning, a particularly fine one, climbed to the summit where a stone platform and a mast were being erected to mark the Survey's actual point of observation. The panorama, which to Everest would by then have been commonplace, took Vigne's breath away.

> I can never forget the glorious view of the snowy range, some sixty or seventy miles from us, as the morning broke over the sacred peaks of Jamnutri and Gangutri ... The entire range of the Himalayas – upon whose most elevated pinnacles the rose-coloured light seemed to pause before it ventured into the yet gloomy atmosphere to the south – was extended from west to east as far as the eye could reach, rearing itself high and magnificently above the great valleys at its base like the turbulent billows of an inland sea.

Vigne ventured heights of '20–25,000 feet' for those magnificent pinnacles; but how high the Himalayas really were remained a matter of heated debate in the 1830s. The debate, if anything, had intensified. Earlier in the century surveyors in India had simply tried to prove that the Himalayas were higher than the Andes and therefore the loftiest mountain range yet discovered. This had been Henry Colebrooke's contention in that memoir on Himalayan observations which had been so rubbished by the *Quarterly Review*. But now, as this contention gradually won acceptance, the question of how much higher they were – and, more especially, which was the highest of all – became matters of much wider interest.

By an odd coincidence, a copy of the *Quarterly Review* which had so roundly discredited Henry Colebrooke's claims was a mute witness to their vindication. It had been addressed to William Webb, Robert Colebrooke's one-time assistant, who in 1819 had returned to the mountains to pursue the headwaters of the Ganges into their deepest recesses. Tracking him up the cliffs and along the parapets of one of the world's hairiest trails, the much post-marked *Review* had finally caught up with him, along with some badly needed provisions, when he was encamped at Kedarnath, a bleak spot 12,000 feet above sea-level where a small stone temple sanctified the source of one of the sacred river's main feeders.

The river, an icy rill, here issued from a chaos of glacier and moraine. That glaciers were commonplace in the Himalayas, despite the mountains being only thirty degrees from the equator, might have caused the *Review*'s critic to hesitate. So would the skyline. While Webb, perched on a boulder, read about how convincing proof had yet to be given that any Himalayan peak was higher than those in the Andes, a cluster of six snowy giants, each superior to Ecuador's Chimborazo, peered incuriously over his shoulder.

Webb, moreover, was reasonably confident of the altitudes which he had just assigned to these giants. By 1820 surveyors were at last getting to grips with the mountains. Instead of long-range sightings from the plains, Webb and others had pushed up the headwaters of the Ganges and Jumna rivers to brush round the flanks of some of the main Himalayan peaks and even pass beyond them onto the Tibetan plateau. They had new ways of assessing altitudes, new instruments for measuring the effects of altitude, and new ideas of how this information could be used to advantage.

The breakthrough had come courtesy of the Anglo–Nepali, or Gurkha, War of 1814–15. Militarily it had been one of British India's less successful aggressions. Kathmandu had proved to be beyond the reach of British arms and the Gurkhas had given such a good account of themselves that their enlistment as mercenaries was reckoned more worthwhile than their submission as feudatories. But as well as acquiring Gurkha recruits, the British had detached certain territories which the Gurkhas, themselves comparative newcomers to dominion, had only recently conquered. These included two adjacent districts, Garhwal and Kumaon, in what had been western Nepal. Besides containing the headwaters of the Jumna and the Ganges, both districts comprised a complete cross-section of the Himalayas – from the dusty outer foothills of the Siwaliks to the Alpine slopes of future hill-stations like Mussoorie, to the rock-strewn canyons and snowy peaks of the main range, and on to the high dry passes into Tibet.

As usual, these new territories demanded new surveys. As well as the sources of India's greatest rivers, 'the heights and distances of the snowy peaks are now within the reach of British research and enterprise,' noted a directive from the Governor-General's office. 'These are objects becoming the attention of an enlightened government.' No longer purely a matter of scientific curiosity, ascertaining the heights of the mountains was now official policy. To Webb was given the responsibility of surveying Kumaon, while Captain John Hodgson, another officer with some experience of the region, was awarded Garhwal. Both had taken the field in early 1816.

Hodgson, who knew something of Lambton's work, realised the importance of starting with a measured base-line. When no suitable site could be found to the south of the outermost Siwalik hills, he made for the Dun, more a broad strath than a valley, which lies at the foot of the ridge on which Mussoorie and Hathipaon now stand. There he again sought a level stretch of ground, but was this time frustrated by the combination of tall elephant-grass, which would not burn so early in the year, and extensive forest, which he lacked the means to clear. Webb was having similar problems. In the end neither secured the ground measurement reckoned so essential for a trigonometrical survey.

Hodgson, however, convinced himself that a long base between two intervisible points whose distance apart could be deduced from their respective latitudes as determined by zenith observations would serve just as well. This was how meridional arcs had been measured by, for instance, Lambton when he began his survey from Madras; and in the almost south–north alignment of over sixty miles between a point in the plains near Saharanpur and The Chur mountain beyond the Dun, Hodgson thought he had the perfect line. Both places commanded fine views of the snowy peaks; and he had soon ensconced himself atop The Chur to take the necessary zenith readings for latitude and, with rather more relish, to train his theodolite on the peaks.

At the time the Great Arc was making painfully slow progress through Hyderabad territory; Everest had yet to join it and Lambton was unable to contemplate its extension beyond Agra. With no chance of the Arc ever reaching the Himalayas, Hodgson and Webb saw their work as transcending its mountain context. Incorrigible optimists both, they believed not only that they would be able to establish the co-ordinates of all the more obvious peaks but that these peaks could then provide 'fixes', as sure as that of the pole star, by which locations throughout the northern sector of the Gangetic plain could be precisely determined. For 'it cannot be denied,' wrote Hodgson, 'that when their [the snowy peaks'] latitudes and longitudes are known, the geographical position of any place from whence one or more of them are visible may be determined with ease and accuracy.' Webb actually detailed a number of case studies to show how this might be done. And both men fervently believed that this simple method held the key to the problem of surveying in the plains of 'Hindustan'. If the Himalayan peaks extended for a thousand miles from west to east, and if they were clearly distinguishable for at least

150 miles to the south, then a vast belt of territory whose murky atmosphere was considered impenetrable for trigonometrical surveying might by this simple method be speckled with the precisely-known locations which a trig survey was designed to supply.

In other words, Hodgson and Webb were turning the problem on its head. Instead of trying to ascertain the heights and locations of the peaks by observations from the plains, they aimed to make it possible to ascertain the heights and locations of places in the plains by observations to the peaks. All they had to do was to pinpoint the peaks, an exercise which promised to be much easier now that they could get to close quarters with some of them. For proximity would afford bigger, and so better, angles for both horizontal and vertical observations, as well as greater definition of the summit to be observed (and so easier identification of it from other observation points) and less distortion from refraction.

They were further encouraged by having theodolites and zenith sectors superior to those available to their predecessors in the mountains, and by having the option of barometers, which instruments had occasionally been tested by Lambton but never yet employed in the mountains. A barometer indicates atmospheric pressure; and two barometers, read simultaneously, reveal the difference in atmospheric pressure between their different locations. By factoring in the temperature at the time and place of each reading, the difference in altitude at which the readings have been taken can also be deduced, since pressure increases with altitude. Obviously this was not much use for measuring the snowy peaks; you had to get your barometer to the top first, and no sane person yet even contemplated shinning up mountains twice as high as The Chur. But it did promise to solve the problem of not knowing the height of the place from which one was observing such peaks. If, for instance, barometrical readings were taken on top of The Chur and other readings were taken simultaneously at sea-level, say in Calcutta, Hodgson could, and did, work out the height of his observation post on The Chur.

There were several other variables which had to be taken into account in such comparisons, but it was the barometers themselves which presented the greatest challenge. Before the 1843 invention of the aneroid (which measures the air pressure by its effect on the surface of a contained vacuum) the barometer relied on the volatile properties of mercury and closely resembled a giant thermometer. A tripod was needed to hold the thing upright, and the glass tube, three feet long, which contained the mercury posed an almost insurmountable carriage problem. Spare tubes were recommended but, as Hodgson found, those not already broken before they reached the mountains soon succumbed to the knocks and tumbles which were an inescapable hazard of Himalayan portage. The last of his tubes was broken on The Chur at the very beginning of his survey. Although some replacements did eventually reach him intact, it proved almost impossible to fill them. The mercury had to be boiled so as to exclude air bubbles, and boiling was as sure to shatter the glass as bashing it.

'Those only,' says an 1850s manual on Indian surveying, 'who have had any practical experience with such delicate and expensive instruments as Mercurial Mountain Barometers can be fully aware of the disappointments met with in a country like this.' Indeed, barometers were so prone to breakage that almost any alternative method was reckoned to be worth a try.

The manual in question recommended a simpler 'thermometrical' solution, which was precisely that to which Hodgson and Webb duly had recourse. All that was needed in this case was a pocket-size thermometer and a kettle. You boiled the kettle and when, as Hodgson put it, the water 'reached full ebullition', you took its temperature. Since the boiling point decreases as the altitude increases, reference to a simple conversion table would give your height. Accuracy depended on very careful reading of a well-calibrated thermometer, and could not be counted on to within less than a few hundred feet. It could also be a problem getting your kettle to boil. Snow and ice, often the only form of water available at high altitudes, took time to melt, let alone 'reach ebullition'. Meanwhile the high winds common to exposed locations toyed with the fire, blowing it either sideways, out, or sometimes away. Nevertheless it is curious that this comparatively simple test seems to have been very little used until Hodgson and Webb adopted it as standard practice.

In the summer of 1816 Hodgson worked north, triangulating where possible, perambulating where not, and planting potatoes as he went. (The potato had been rightly identified by the then Governor-General as having an important contribution to make to Himalayan diet and economy.) But, 'having some doubts as to the precise latitude of my grand station on The Chur on which everything depends', Hodgson returned there in October and again in 1818 after an adventurous foray to the most distant of the Ganges' sources. Near the top of The Chur he had constructed a thirty-foot pyramid from which, as well as from Saharanpur in the plains and from Surkananda to the east of Mussoorie, he plotted the angles of some fourteen major peaks and many more minor ones. In fact The Chur for Hodgson, as later for Everest, became a second home. He wrote of it with feeling – although not *in situ*, since the ink invariably froze to the pen.

At a place like the summit of this vast mountain no one who has not resided in such a stormy region can have an idea of the violence of the wind, and the suffering of an observer by night from the cold ... On the 10th of October, water instantly froze when poured out at 9 in the morning, tho' the sun shone out, the thermometer being then at 31°. Judge then what it is by night, accompanied by a wind which peels the skin from the face, and blows with a violence which seems to shake the very ground. I had a tent to protect the instrument [his theodolite] but at the time of observation the wind rushes in and shakes the instrument, and blows out the lights and creates confusion, and people holding the tent to prevent its being carried away are apt to touch or shake the stand [of the instrument] so that I found it impossible to keep the adjustments in order from night to night.

Nor was this the worst of Hodgson's problems. The latitudes of his Saharanpur-Chur base as established by zenith observations simply could not be reconciled with those computed by triangulation from a third point. Trying other points from which to complete the triangle made no difference. And the discrepancy remained when his colleague and friend James Herbert carried out a check. It was said that if experienced observers, taking all possible precautions, found themselves confronting an anomaly for which they could not account, they were probably 'on the verge of some important discovery'. Hodgson quoted this maxim with approval and was perfectly candid about the discrepancy, hoping that others might someday be able to identify what it was that he had discovered.

In fact, of course, they already had. But Webb, Hodgson and Herbert seem to have been totally unaware that mountain masses exercise an attraction over the plumb-line (so critical in zenith readings). Nor were they aware that, as per Lambton's conjectures, the earth's density might have a completely contrary effect. Moreover, they were in the worst possible place for this distortion. A geodetic paper on the subject would later confirm that the plumb-line deflection amounted to fifteen seconds of a degree at Saharanpur and at least thirty-six seconds at The Chur. The resultant discrepancy would be something like a third of a mile in the sixty-one miles of Hodgson's base, a quite unacceptable ratio of error which, according to the Survey's historian, 'rendered all Hodgson's care and labour of no avail whatsoever'.

To resolve the problem, Hodgson did in the end revert to his original plan of a ground measurement. But, after long exposure on the hill tops, his health was no longer up to the task and it therefore fell to his assistant and successor. To James Herbert in 1819 belongs the honour of measuring the first Himalayan base-line.

Herbert was clearly a resourceful individual. Although he had no experience of base-line measurement, had never even witnessed it, and lacked much of the necessary equipment, he entered into the business with enthusiasm. His only complaint was that he had no assistant. Lambton reckoned that at least three Survey officers were essential to oversee such an operation and Everest, for his Calcutta line, would muster no fewer than ten officers. Herbert was alone except for his Indian staff, none of whom knew the procedures any better than he, and none of whom he could trust to take a reading. How many times he had to scuttle from one end of each measurement to the other he does not say, but it must have been many.

A site was found in the Dun which, though on a slope, was as near flat as anywhere in Garhwal and Kumaon, and the necessary vegetation clearance was soon underway. Meanwhile Herbert wrestled with the means of measurement. He had a chain but no coffers, no tripods and no elevating screws with which to level the tripods. Assembling and making these things would be the work of months. Reserving the chain for comparison purposes, therefore, he decided to use rods. A well-seasoned roof joist of Himalayan cedar was removed from a derelict house and sawn into four lengths, each of twenty-five feet. William Roy had tried rods on his Hounslow base, trussing them to reduce bowing and fitting the ends with ivory markers. Herbert used brass buttons for the critical contacts but, with his rods being only 1¾ inches square and trussing being out of the question, he was much concerned how to prevent them from bowing. In the end he devised a retractable stand on which the rods would rest, and ordered thirty-seven of these, one being required every six feet. As a visual check on bowing he also tensioned a thin brass wire which ran the hundred-foot length of his rods in a specially cut groove.

A bench incorporating a brass standard was designed for the daily check on the expansion or contraction of each rod, and various markers, pickets and flagstaffs were also improvised. All in all, and considering that everything except the rods themselves had to be made out of unseasoned pine, Herbert was not displeased. In the final measurement of about four miles he reckoned that the error could not exceed two feet, 'an uncertainty which will only affect the distances of the remotest peaks by about sixty or seventy feet'.

Once the carpenters had finished, the whole operation took no more than a month. It then remained only to connect the base by triangulation to Hodgson's problematic Saharanpur-Chur-Surkananda triangle. This done and the doubtful zenith readings of the triangle corrected, the observations previously conducted to all the snowy peaks could be adjusted and reduced to a table entitled 'Latitudes, Longitudes and Elevations of principal Peaks and Stations in the Survey'. It was published in 1822, just as an apoplectic Everest was laying out his Sironj base-line on the Great Arc.

Of the forty-six Himalayan peaks in the Garhwal-Kumaon region which Hodgson and Herbert had located and listed, only fourteen were less than 20,000 feet above sea-level. Five were over 23,000 feet, and three of these giants, which Hodgson had numbered as 'A1', 'A2' and 'A3', were far enough east to have also come within Webb's Garhwal survey, with which they provided a useful link. Webb too had remarked the height of this threesome and had produced very similar values. At Hodgson's 25,749 feet, 'A2' proved the biggest of the three; and as Herbert discreetly put it in a note printed so small as to be barely legible, 'A2 is thus, so far as our knowledge extends, the highest mountain in the world.'

Reviewing the Garhwal and Kumaon surveys, Andrew Waugh, Everest's successor as Surveyor-General, would find them 'highly creditable to the scientific ability of the officers employed', although he greatly regretted the 'inartistical' nature of their maps. Everest himself applauded Herbert's efforts to determine a base-line and rated the Garhwal survey 'quite sufficiently accurate for geographical purposes'. Where not duplicated by his own triangulations, it would be incorporated into the Great Trigonometrical Survey. He was less happy with Webb's effort in Kumaon. Lacking a base-line, he thought it 'little better than what is called a wheel and compass survey'.

Webb's observations had, though, been good enough to win a rather oblique retraction from the *Quarterly Review*. Tucked away in an 1820 notice of a recent paper in French 'On the Height of India's Mountains', the *Review*'s contributor cited findings lately received from Webb. Evidently Webb, incensed by what he had read at Kedarnath in the earlier *Review*, had quickly set up his barometer. The readings, forwarded to London, convinced the reviewer that Kedarnath stood at 12,000 feet above sea-level and that a nearby pass into Tibet was over 16,000 feet. Both places were snow-free at the time. According to Alexander von Humboldt, the distinguished author of the paper under review, who had made extensive observations in both South America and Europe, such altitudes should be well above the level of perpetual snow and must therefore be exaggerated. Earlier the *Review* had also used this argument. Now it conceded that snow-lines were not necessarily consistent and that its conclusions had been 'as erroneous as those of the Baron de Humboldt'. Moreover, if the heights of such observation points were correct, then so probably were those of the peaks observed from them.

Webb's sightings, back in 1808, of Dhaulagiri from south

of the Nepal border were still suspect. The distance had been too great, the angle too low, refraction too uncertain, and his instrument too primitive. Even Webb now agreed. Crucially, the distance between the places from which he observed the mountain and calculated its position was also suspect, in that it had been deduced from zenith observations just like Hodgson's at Saharanpur and The Chur. Until the Central Himalayas in Nepal could be approached as closely as the Kumaon and Garhwal Himalayas, or until such time as they could be observed from points in the plains whose locations were more precisely known, such long-range observations were reckoned a waste of time.

But the heights given in Herbert's 1822 listing for Garhwal looked good. They had been grounded on his base-line, however crude, and they had been carefully observed at comparatively close quarters from triangulated positions whose altitudes had sometimes been verified barometrically. 'A2' therefore went unchallenged as the world's highest peak. Curiously, it also went unnamed. Herbert's list included a Mount Moira (after Governor-General Lord Moira) and Mounts St Patrick and St George, all clustered around Gangotri. They had been so christened by Hodgson, either in an onset of patriotic fervour or in a gesture of Anglo-Christian bravado. Contemporaries disapproved and the names have since been expunged from the map.

But for 'A2', neither Herbert nor Webb had proposed a name. Observing it from a distance of about fifty miles, they had been unable to ascertain whether there was already a local name for it. Moreover, they probably felt disinclined to suggest one. For were it to remain the world's highest, this honour would almost certainly be contested by their superiors.

Such anxieties, however, proved groundless. In the course of time it emerged that it did indeed have a local name – Nanda Devi – and that it was not the world's highest. But Nanda Devi is, still, the highest mountain in India (as opposed to Nepal, Tibet and Pakistan-controlled Kashmir); and at 25,645 feet its official height today is only a hundred feet less than that given by Herbert.

NINE

Through the Haze of Hindustan

The outermost limit of the Himalayas in Garhwal is defined by the Siwalik hills, a brown sierra of modest height which intervenes between Mussoorie's alpine slopes and the baking Indian plains. Here, on the morning of 12 November 1833, a small army might have been seen breasting the ridge and, with the dew still on their boots, slithering down parched and gravel-strewn gullies, past Saharanpur whence Hodgson had begun his Garhwal survey, and on into the dusty immensity of the Gangetic plain. After an interlude of eight years, George Everest was taking the field again; and with its Superintendent back at the helm, the Great Trigonometrical Survey was launching itself into the dreaded haze of Hindustan. Departing Hathipaon with two assistants, three sub-assistants, four elephants, forty-two camels, thirty horses and 'about 700 natives' – in that order – Everest was about to address what he considered the most difficult terrain ever to be triangulated, let alone trigonometrically surveyed to the exacting standards of the Great Arc.

During the previous season he had done little more than assess the challenge. From his well publicised base-line measurement in Calcutta, he had journeyed east to Sironj, where the Arc had been abandoned seven years previously, and had then reconnoitred north through Gwalior, Agra and Delhi along the line which it now must follow. On the way, he had met up with Olliver, Rossenrode and some of the newly recruited assistants. All were already marching and counter-marching in search of hills, mounds, barrows, buildings, anything with a view which might serve as a trig station. But the season was exceptionally dry, the atmosphere a 'pea soup', and the heat soon became like nothing Everest had ever experienced. One of his parties was shot at, another robbed. All reported numerous fever cases.

Rossenrode, currently rated the most reliable of veterans, had managed well in the more undulating country south of Agra; but the dependable Olliver had let Everest down on the section thence to Mathura (Muttra); and a man called Boileau, lately transferred from topographical surveying, had been an unmitigated disaster wherever he was deployed. Everest had hoped to complete the selection of stations by May 1833. In fact it had barely begun.

Continuing north, he had found that whereas up to Delhi he might get away with building towers on only one side of the Arc, from there onwards every station would require a tower. According to the co-ordinates worked out for places in the plains by Herbert and Webb using their peaks as 'fixes', the 78-degree meridian passed straight up the 'two rivers' region between the Jumna and the Ganges. Known in Hindi as the Doab, this was a congested flat even by the horizon-choked standards of the Hindustan plain. Up to Delhi, the Arc could advance with one 'foot' on rising ground to the west of the Jumna. Thereafter it would have to cross to the eastern bank and plunge both 'feet' into a cloudy murk whose floor was strewn with villages and encumbered with trees.

Fuming over the season having been such an abject failure, Everest had then retired for the monsoon to his new property in the hills. It was his first visit to Hathipaon. There were leaks to be fixed, workshops to be built and accommodation to be organised. As the monsoon battered the ridge, Everest's spirits revived. In a make-or-break gesture, he vowed to take the field personally at the head of his entire establishment as soon as the rains were over. Hence the cavalcade which crossed the Siwaliks in November 1833.

They headed straight for Mathura, halfway between Delhi and Agra. There and elsewhere Everest had left instructions for the stockpiling of timber, bamboo poles, ropes, blocks and pulleys. These items were now divided up into numerous cart, camel and elephant loads and despatched to depots along the line of the Arc.

Triangulating such impossible terrain involved different procedures to those followed by Lambton when *droog*-hopping across Mysore or by Everest himself when working up through central India. A simple linear progression in which an advance party chose the best hills while the Superintendent, following along behind with the Great Theodolite, took the angles, was out of the question. 'I was about to make my first essay in a new career,' wrote Everest, 'wherein all my former experience would avail me but little.'

Except where some existing eminence invited attention, stations in the plains could be located almost anywhere. The skill lay in finding the least objectionable position, somewhere sufficiently distant from, and at an appropriate angle to, other stations whose lights might eventually be sighted through the dense atmosphere after a minimum in the way of tree-felling, house-demolition and ground-levelling. But just how much clearance would be necessary depended on where the sight-line would actually fall. Would lopping a single branch solve the problem? Or would a whole village have to be moved? With sight-lines grazing the ground for up to thirty miles, it was hard to tell.

It was therefore essential to predetermine the direction of each line mathematically. A chain of small triangles conducted between the two positions could provide this information, but the procedure was time-consuming and often abortive. Better was the system developed by Everest which he called 'ray-tracing'. For this a party advanced along the supposed line with a perambulator and compass, traversing to left or right at exactly ninety degrees in order to work round any obstructions, and likewise in order to home in on the final position. From the angle followed, the traverses involved and the distances recorded, the true bearing of the sight object could be calculated. Telescopes and heliotropes could then be trained accordingly and the 'ray' duly cleared of obstructions. It was an extremely trying business. Yet the cost of a mistake, like erecting a tower where it would not serve its purpose, was unthinkable.

To avoid such a catastrophe, stations were also tested by a series of preliminary experimental or 'approximate' triangulations made from temporary structures. This work could be going ahead on different sections of the Arc simultaneously. Likewise, once the preliminary triangulation was completed, base-lines and astronomical operations could be undertaken while the masonry towers were being built for the final triangulation. The whole four-hundred-mile sector of the Arc north of Sironj was thus treated as one, and progress was measured not in miles advanced but in operations completed.

The selection of stations and their preliminary triangulation having largely failed in the previous season, their completion was the main task during the dry weather of 1833–4. Leaving others to finalise Rossenrode's work up from Sironj, Everest commenced operations at Fatehpur Sikri, thirty miles from Agra. This was on the easier section south of Delhi, and Fatehpur Sikri actually stood on a low hill. It was here in the 1570s that the Mughal Emperor Akbar had re-sited his capital. More a palatial set than a city, Fatehpur Sikri's pristine halls and courtyards, all in the same dull red sandstone, had quickly palled on the Emperor. He abandoned the place in the 1580s and, after subduing most of India, was eventually buried beneath another noble pile of sandstone on the outskirts of nearby Agra.

To the flat and domeless rooftop of this latter mausoleum Everest now ordered one of his signal teams preparatory to reconnecting the dead city of Fatehpur Sikri with its dead emperor at Agra. The tomb had been damaged by British military operations in 1803 and was presumably reckoned less sacrosanct than that of Akbar's grandson, Shah Jehan, whose Taj Mahal in the heart of Agra was also well within Everest's range of vision. As with other notable landmarks, the position of the Taj was duly observed and for the first time precisely recorded. But tempting though it may have been, Everest refrained from scaling its great white dome. Mercifully, that moist 'tear on the face of eternity', as Rabindranath Tagore would call it, never suffered the indignity of being dabbed at by a Survey flag.

On Akbar's tomb the task of raising and roping the twenty-two-foot flagstaff fell to the much-maligned Captain Alexander Boileau. Boileau had been making the most of his stay in Agra. The city's Executive Engineer happened to be his brother, so there had been no problem about obtaining authority to remove a pillar from one of the tomb's crowning cupolas when it interfered with his sight-line. Between such acts of casual vandalism, he had also taken the opportunity to propose to Charlotte, the sister of his brother's wife. When Everest moved on, the pair would hastily marry before Boileau himself was shunted north up the Arc.

Over at Fatehpur Sikri, a platform had been prepared and a stone marker to record the point of observation had been sunk deep in the ground. Everest was expected any day. 'The evening of my arrival I shall light two large [bon]fires, for which please keep a look-out,' he wrote to Boileau. 'I wish you to burn a dozen blue lights at intervals of a quarter of an hour ... If you see the double fires, allow half an hour to expire after their first blaze before you burn your first blue light ... If you can lay down the approximate position of [Akbar's tomb] it will assist me ... I remain your most obedient servant, George Everest.'

Convention alone dictated the closing sentence; Everest was no one's obedient servant, least of all Boileau's. The system of bonfires and the use of flares, rather than terracotta lamps, was also now standard. By following Everest's instructions carefully, the flares – like large fireworks except that each was sealed into a sheep's bladder – could be made up locally. The recipe involved 739 parts ('sulphur 136 parts; nitre 544; arsenic 32; indigo 20', etc., etc.), and was not susceptible to improvisation. Any adulteration and the flare would not light, any variation and it might explode; and much as Everest enjoyed a shower of falling stars, they were 'extremely inconvenient for observation'. Finally, each flare should weigh three pounds, so that '160 will be the load for a camel'.

Needless to say, Boileau's flares performed dismally. They deluged his men with lava and spluttered sparks to useless effect. A month later the wretched Boileau was reported absent without leave. Perhaps he was on honeymoon. But as Everest noted with unconscious irony, 'it is not the first time.' Boileau was discharged.

Everest continued north. As the country levelled out, new sighting devices were tried. Flagstaffs gradually gave way to masts (as objects to observe), and platforms to scaffolds (as the places whence to observe them). A mast was typically seventy feet high and consisted of a core post, as long as possible and deeply embedded in the ground, around the top of which long bamboo poles were firmly lashed. The upper extremities of these poles being far higher than the post, they were themselves tied so as to form a trunk. Another section consisting of more poles was then lashed round it – and so on, like a giant fishing-rod but with the whole thing being as liberally secured with stays and guys as a telecom mast.

At the very top was fixed a pulley by which a single bamboo pole of some forty feet could be hauled up in a horizontal position. This was in effect a boom, to one end of which the flare could be attached while from the other end dangled a long rope. When the flare was lit, the boom would be smartly hoisted to the top of the mast by the pulley, and then manoeuvred into the vertical by pulling on the dangling rope, 'thus supplying,' according to a triumphant Everest, 'a brilliant blue light at upwards of 90 feet above the surface of the ground'.

Meanwhile many miles away the observer, usually Everest himself, was anxiously peering down the telescope of his theodolite from a platform at the top of a more substantial structure. The theodolite itself stood on its stand, which was mounted on a circular table bolted to the top of a thirty-five-foot spar of seasoned timber. At least five feet of this veritable tree-trunk were buried in the ground, and at twenty feet up, an iron collar afforded attachments for stout stays and 'antagonising struts'. The idea was that the structure, and so the instrument, would be totally rigid and impervious to vibration. That meant that the observer had somehow to reach it and operate it from a separate structure. A scaffolding with ladders and a tented observation platform was constructed around the mast but completely independent of it.

Such, at least, was the theory, although in practice it was often found that because of the wind, the mast did indeed move. More bamboos were then introduced to lash it to the posts of the scaffolding, thereby rather compromising the principle and obliging everyone to tiptoe shoeless up the ladders. While observations were actually in progress, no one was permitted to so much as move on the scaffolding. Only Everest and someone to hold the lamp by which he took the readings could stand. The rest – the assistant with the angle-book, the lampman with the oil, the instrument attendant with his duster – had to squat on the floor and stir not a muscle.

Supposing the theodolite to be a camera, the scene must have resembled that of a night shoot on a movie set. In the ingenious use of bamboo scaffolding the Bombay film industry probably surpasses the Survey; but in the hush of expectation as Everest climbed to his platform and 'stood to the instrument', in the cry of 'Lights!' and in the endless retakes, any movie-maker would have felt at home. Save for a canvas chair and a plastic eyeshade, Everest could have been an eminent director. His role was that of orchestrating a vast production in which his various 'crews' were expected to heed his every command without necessarily comprehending his vision. To one who saw beauty in every angle and truth in every equation, scientific probity was tantamount to artistic integrity. As the mastermind behind the whole enterprise, Everest felt entitled to a deference which transcended rank and bordered on reverence. Creative genius was at work; an enormous expenditure had to be justified, a perfectionist's reputation upheld, and a monumental ego sustained.

If the Great Arc was the star of this production, all India was its set. Outside of the Survey, Everest had few friends and no interests. A break from fieldwork was just an opportunity for catching up on the backlog of computation and correspondence. His dedication was as absolute and undisputed as he proclaimed it to be, repeatedly. Yet the unreasonable demands, the histrionic outbursts and the deeply offensive language could not but rankle.

'You are mismanaging sadly,' he told Henry Keelan, one of his new sub-assistants; 'when instructed to turn your heliotrope to Bahin, you turned it to Pahera . . . you might as well turn it to the moon.' A wilting Keelan proved just as irregular with his night flares. Again he was castigated, but this time he countered by explaining that he did not own a watch. 'No decent person is ever without a watch,' thundered Everest. 'You ought to be ashamed . . . You might just as well say that you have no coat or no shoes or no hat.' Still watchless, Keelan continued to get his timings wrong and to drive Everest to despair. 'You are evidently one of those uncertain persons in whom no sort of confidence ever can be placed. Ask yourself what use a person can be who commits blunders so often . . . Sometimes you are too lazy to get up in the morning. At other times you make intervals [between flares] of 32 minutes instead of 16. At other times you break the pole. In short you never succeed except by the merest chance.'

Keelan and another new recruit, Charles Dove ('where there is a will, there is a way, Mr Dove'), were singled out for particular censure. Both were recalled, Keelan in disgrace, 'the faint-hearted Dove' after being struck by a falling mast. More to Everest's taste were two young Lieutenants who had joined the Survey in 1832. Andrew Scott Waugh and Thomas Renny-Tailyour had received some training with the Irish Survey. They were promising mathematicians and 'tasty draughtsmen'. Best of all, they were officers and gentlemen, descendants of landed Scots gentry and recipients of a good education. Everest had never approved of Lambton's 'mestizos', and after an uneasy induction with, but not under, William Rossenrode, each of the Lieutenants was promoted above Rossenrode as full assistant, rather than sub-assistant. They were then given command of one of the meridional series running north from Olliver's Calcutta Longitudinal Series. But as men like Keelan and Dove fell by the wayside, Waugh and Renny would be increasingly summoned to the Great Arc and would perform much of the final triangulation.

With the preliminary triangulation approaching the ancient capital of Delhi, it was Rossenrode who suffered another tumble from grace. The city's most obvious natural feature was The Ridge. It flanks Old Delhi on the north-west and would become hallowed ground for the British when, twenty-three years later, they there staged a do-or-die struggle against the forces of Indian resurgence in what they called the Indian Mutiny.

Rossenrode had been sent to find a suitable site on The Ridge to serve as the next station. He soon discovered that, then as now, Delhi's atmosphere was amongst the most pol-luted in the world. Thirty miles away, Everest, although he himself could make out very little, seemed to imagine that Rossenrode, in the midst of the haze, had only to pick out his signals and settle on a site. Why was he taking so long? If he could not see the signals, he had only to shift his position. 'You are wearing me to fiddlestrings about this Delhi ray . . . If you do not take some pains, you will never succeed and I may be detained here for the next six years. It is pleasant enough for you, I dare say, near a grand cozy city, but for me and all about me it is a great nuisance, I assure you.'

Like Keelan and Dove, Rossenrode was recalled. Immediately, as if by design, the fog cleared enough for the signals to be seen. Crowing over this disgrace of one of his senior sub-assistants, Everest hastened into town. There he quickly discovered that the sightings were worthless and that the buildings on The Ridge were mostly too unstable to serve as a station. He lit upon an old mosque but eventually adopted a domed building which looks to have been that originally chosen by Rossenrode. Known as the shrine of 'Pir Ghalib' or 'Pir Ghyb', it would later surface in guidebooks as an 'ancient observatory'. Twentieth-century visitors, oblivious of the Great Arc and all that it involved, had evidently identified as pre-British both the hole drilled by the Survey in the dome and the corresponding hole and marker directly beneath it in the floor of the building. This was the standard method of ensuring that the instrument on the roof was precisely plumbed above the marker. It would be replicated in the custom-built towers whose design Everest already had on the drawing board.

A month later and now within sight of the mountains, Everest turned on his most senior sub-assistant. Joseph Olliver, one of Lambton's Madras protégés, had been with the Survey for nearly thirty years. He had accompanied Everest in the jungles of Hyderabad, commanded the Calcutta Series, and had become Everest's most consistently successful triangulator. He had also, like Rossenrode, been followed into the Survey's employ by three of his sons. Described as being 'of a retiring disposition' – and certainly an uncomplaining one –

he had now served as Everest's foil for a decade and a half. He deserved better than the outburst which followed an unsuccessful night of flare-burning.

I dare not put a blue light into the hands of any of you. You seem to think they grow like grass, and that all you have to do is put them at the top of the pole and set fire to them, just as you would to a whisp of grass. I suppose the only way is for me to ... leave you to recover your senses, for it seems that you will not abide by my orders, but – pell-mell, helter-skelter, foul or fair – away go to damnation and destruction the only means we have of getting through our work.

You all seem to me to be right stark staring mad. Never was a worse evening ... I could not see five miles in any direction. The sun was obscured at 4 o'clock, and by 5 there was not a vestige of him, and that is the kind of atmosphere in which you choose to burn blue lights. I have superseded Mr Dove ... I have sent out Mr Keelan ... and unless I receive some assurance that you will not play the fool in the like manner again, I shall certainly adopt equally strong measures to you.

An unamused Olliver no doubt shuffled away into his retiring disposition.

As the cool dry season gave way to the hot blasts of April, the preliminary triangulation was carried up into the dusty Siwaliks. Hoisting the Arc over these outermost sierras and down into the Dun entailed finding two intervisible stations on the crest of the range. It should have been easy. But following weeks of no hills, now there were too many. Amidst reports of Everest's men fleeing from tigers and being chased by rogue elephants, six positions were in turn occupied and abandoned. Nor were the seventh and eighth ideal. In a blatant example of surveyors manipulating the geography they were so intent on measuring, intervisibility was secured by reducing the height of an intervening peak. With crowbars and sledge-hammers, twenty feet of rock were pruned from the profile of the Siwaliks.

Everest was in no mood to be troubled by scruple. Cooler climes and clearer vistas beckoned. Hathipaon was now visible across the Dun; respectable peaks, like The Chur, loomed invitingly beyond; and along the furthest horizon marched the serried snow-caps of the Great Himalaya. With the end in sight, he penned a triumphant report in which he announced that there was 'no instance on record of a symmetrical series of triangles having been carried over a country similarly cir-cumstanced'. Every station for the final triangulation had been selected, the required height for every tower ascertained, and at least two angles of every triangle approximately measured. It was, in short, 'an unbounded success'. And for it Everest, as usual, took unbounded credit.

Fourteen towers would be needed for the final triangulation, at a cost of about two thousand rupees apiece. Basically each was a very solid rectangular version, in brick or stone, of the mast and scaffolding used for the preliminary observations. Again the instrument table was mounted on a pillar isolated from the observation platform. The table was centred over a shaft down which the plumb-line could pass to the marker stone in the ground below, and the platform was always on the roof of the topmost storey beneath a canvas awning. Heights varied from forty to sixty feet, and so did the architec-ture. To judge from those which survive, the most popular model owed something to the bell-towers of Tuscany.

Access was by ladder up the outside. Officers of the Great Trigonometrical Survey were not accustomed to such luxuries as stairs, says Everest. To haul to the top the half-ton Great Theodolite, now rebuilt and about to re-enter service, a crane was mounted on the topmost platform. Unfortunately it was found to restrict visibility. A lesser derrick had to be constructed to dismantle the crane, the derrick being of a size to be then manhandled to the ground.

The building work was to be undertaken by different engineers attached to the various British military establishments along the line of the Arc. Plans were sent to them, detailed instructions given, and a lively correspondence was generated. But it would be two years before they were completed. Meanwhile the Arc's conclusion could be anticipated by measuring the final base-line, selecting terminal stations in the Himalayas, and preparing for the delicious certainties of the final computations. That, after thirty years, the longest and most ambitious meridional arc in the world would be successfully completed was no longer in doubt. Whether Everest would be there to claim the unbounded credit for it was a different matter.

TEN

Et in Arcadia

S lumped on the grass outside Hathipaon, watching the mist scud over the ridge, I had got to thinking about the extraordinary irony of it all. How had something as rigorous and predictable as the Great Arc had such unforeseen consequences? It was as if scientific endeavour were subject to a law of *karma* which ordained that every experiment must be productive of its antithesis. Thus a sweat-soaked odyssey conducted across the burning plains of India would uncover the frozen secrets of the highest Himalayas; a measurement intended to discover the curvature of the globe would reveal its greatest irregularity; and a man who dealt in decimals to the sixth place and degrees to a hundredth of a second would find his name attached to the most colossal of mountains.

More extraordinary still, what was by common consent one of the greatest scientific achievements of a science-mad century would go practically uncelebrated. It was very strange. Was science always so capricious, invention ever so perverse? As John Hodgson had noted, it was when the pioneer was apparently confounded that he stood on the threshold of a discovery.

Getting up, I had wandered towards the abyss beyond the terrace on which Hathipaon stands. Four thousand feet below lay spread the Dun. A broad and now open expanse of farmland about thirty miles long by ten wide, it separates the Ganges and Jumna rivers as they emerge from the mountains. Beyond it to the south, the crumbling hill profile was that of the Siwaliks – less a twenty-foot pinnacle; and beyond the Siwaliks stretched interminably the north Indian plain. From the ridge of Hathipaon the great plain was just discernible as a bilious haze pulsing with heat. There lay Delhi and Agra, their minarets and tower-blocks hidden at this range even to the telescope of a thirty-six-inch theodolite.

In contrast, the Dun in the foreground was clear in every detail. It was laid out, indeed, like a map. Immediately below the ridge, Dehra Dun, the main town, sprawled uncertainly outwards from a core of corrugated roofs to the parks, parade-grounds and arboreta of its prestigious academies, fee-paying colleges and government institutions. Amongst the latter I thought I could identify the headquarters of the Survey of India, the organisation over which Everest had presided in his role as Surveyor-General. In the Survey's offices, alongside portraits of all the other Surveyors-General, hangs that leonine likeness taken long after his retirement. There, if you mispronounce his name, they still actually correct you. 'Oh, you must be meaning EVE-rest.' It is more than fifty years since the British left India but at the Survey he is yet remembered, his outbursts fondly quoted and his instruments proudly displayed. Quite apart from his scientific achievements, he is recalled with affection as the man who was responsible for first locating the Survey in the salubrious township of Dehra Dun.

In 1833, soon after Everest's office and instruments had been so laboriously shipped upriver from Calcutta to Hathipaon, the government had begun raising objections. Its Surveyor-General, his staff, and the whole map-making directorate had no business relocating themselves in such a remote and inaccessible eyrie. The Survey's headquarters were supposed to be in Calcutta. There, in the capital, fretting officials and idle presses awaited the surveys and charts which Everest's department was supposed to be churning out. Throughout India new roads were being planned, irrigation canals laid out, and the first railways projected. There were new districts to be pacified, frontiers to be drawn and, most important of all, whole territories to be 'settled' (this being a euphemism for assessing and allocating the agricultural taxes which made India such an attractive country to rule). For all these activities maps were essential, and the Survey of India was there to provide them. It was quite unacceptable that it had decamped to a Himalayan retreat where it could barely be contacted, let alone supervised.

Everest, of course, protested. If he was to perform his dual duties, his headquarters as Surveyor-General would have to be handy for his fieldwork as Superintendent of the Great Trigonometrical Survey. That meant being near the Great Arc. Additionally, given the health hazards to which his men were exposed, headquarters needed to be located somewhere with recuperative potential plus a good monsoon climate, that being the only season when he could devote himself to administrative duties. Hathipaon, he insisted, was both. But the government remained unimpressed, and it was only reluctantly that a final compromise was accepted. While the computational, graphic and administrative core of the Survey of India was to remain in Calcutta under Joshua de Penning, the headquarters staff need only move down from the heights of Hathipaon to offices in the more accessible Dehra Dun. There they have been ever since.

Everest himself resisted even this short removal. As Surveyor-General he was expected to follow his office down to the town, but as Superintendent of the Great Trigonometrical Survey he stayed put at Hathipaon. The workshops on the ridge continued to echo to the sound of hammers and grinders; and whenever operations in the plains were suspended, Logarithm Lodge and Bachelors' Hall overflowed with his henpecked assistants. Everest would continue to run the Great Trigonometrical Survey, if not the Survey of India, as an extension of his domestic arrangements, and in the absence of a family he rejoiced in playing the awesome patriarch to his staff. Government might whinge about overdue maps and critics whisper about the neglect of surveys other than the Great Trigonometrical Survey. Everest knew better; his grid-iron of triangles must come first.

Scanning the Dun westwards, I strained for signs of a place called Arcadia. It should have been four or five miles to the right of the town and on the far side of the Asan rivulet. But with no idea what to look for, I saw only villages and denuded fields, many now fringed with eucalyptus in a pallid apology for the great woods of *sal* and pine which once gave this landscape an elysian appeal. Everest, no doubt, would have been able to direct his gaze straight to the spot. It had been his main reason for buying Hathipaon in the first place. On the strength of Hodgson's observations, he knew that the 78-degree meridian passed through the Dun, and on the strength of Herbert's experience, he knew that only in the Dun was he likely to find the six to seven miles of tolerably level ground necessary for his Himalayan base-line. From the ridge, Hathipaon was sure to command the site, and in 1833, soon after his arrival, he had 'had the good fortune to pitch exactly on the tract which proved in the end to be the most favourable'.

It was not the site used by Herbert. That lay on the other side of Dehra Dun. Everest needed a base-line whose terminal stations would connect with the two trig stations he had established on the Siwaliks and with The Chur, on whose summit he had already established another station. With the heavy compensation bars to position and with ample staff to assist, he planned something much more elaborate than Herbert's base. Extensive clearance was undertaken, bridges were built where the terrain fell away, and stone markers were sunk in the ground at either end. These markers were then enshrined in tumuli and eventually topped with towers. In addition to the measurement of the base, the compensation bars themselves were compared against a brass standard before, during and after the measurement in order to detect any inconsistency. Following several hundred such comparisons, Everest pounced on a possible error which, compounded over the seven miles of the line, might come to 1.6 inches. He was disgusted; it was nearly double the margin of error found on the Calcutta base. In a rare criticism of his beloved bars he queried whether their accuracy was 'commensurate with the increase of complication in the machinery and the expense incurred'. The bars, as he would discover, were nevertheless a lot more reliable than a chain.

The whole Dun base-line operation took from October 1834 till February 1835. It involved nearly all Everest's assistants and sub-assistants – Waugh and Renny, Olliver and Rossenrode, plus three promising newcomers, Peyton, Logan and Armstrong, who between them would later explore the Himalayan potential of the Great Arc. Even the watchless Keelan, now reinstated, was allocated to one of the microscopes on the bars. Beside him worked Radhanath Sickdhar, a twenty-year-old Bengali recruited by de Penning for computational work in Calcutta and since poached by Everest as his number-crunching genius. As well as being the first Indian of rank in the employ of the Great Trigonometrical Survey and its undoubted mathematical star, Sickdhar too would be credited with Himalayan discoveries.

No conclusions could be drawn from the new base-line until it had been connected up by primary triangulation with that at Sironj. This work still awaited the completion of the masonry towers. Meantime instruments were readied and Everest switched his fire from his subordinates to his neighbours. In one of the more bizarre rows to rock British India the original culprit was a bored mule which, escaping one afternoon from the Survey's Dehra Dun compound, entered a neighbouring garden in search of a herbaceous bite. The neighbour, a Lieutenant Henry Kirke who was also the town's Staff Officer, demanded satisfaction. When none was offered, Kirke retaliated: a herd of cattle appeared in the Survey's compound. Some fouling of equipment resulted and several straw-thatch houses were eaten while their inhabitants, Survey employees, were out clearing the base-line. Everest, lately promoted to Major, promptly impounded the cattle, and Kirke demanded their release. When his demand was a second time ignored Kirke had recourse to arms. In an unseemly fracas, a Sergeant and four Privates with fixed bayonets routed the 'compass-wallahs' and reclaimed their herd.

Everest was by now beside himself. Long-winded letters stuttering with rage and innuendo were fired off to the town's commanding officer and to the Adjutant-General in Calcutta. But Kirke gave as good as he got, and with the affair getting out of control it was referred to the Commander-in-Chief.

There, mercifully, the correspondence ends; who came off best is not known. But as a result of this and other run-ins with the civil authorities, Everest's standing suffered. No one questioned his competence or dedication, but his arrogance was cordially detested and his outbursts openly ridiculed. Not all India would remember him fondly, and his achievements would be tarnished in consequence. When the later naming of a peak in his honour proved so controversial, it looks to have had as much to do with the man as with the mountain.

The reputation of his office also suffered. Indeed the Survey was so 'hated', according to one contemporary, that its unpopularity discouraged promising recruits. On the other hand, the sequel to this affair of the mule is also revealing. Just as the saga mysteriously vanishes from the records, so the hatchet seems to have been quietly buried in the Dun. Kirke and Everest subsequently got on well. It was Kirke who would purchase the ground across which Everest had laid out his base-line and, planting it up as a tea garden, would call it Arcadia. The name, reported a gleeful Everest, was 'in commemoration of, and compliment to, the Great Arc!!!', a handsome gesture which he handsomely acknowledged with congratulations to Kirke on his tea bushes as well as those three unwonted exclamation marks. Clearly, those who fell foul of the fiery Major, whether subordinates or colleagues, were not left to nurse their grievances.

Everest's irascibility, like that of generations of other choleric Englishmen in the East, may charitably be explained by his recurrent ill-health. Rebellious bowels lent an urgency to the working day, and malarial rheumatics made nocturnal observations an agony. From England, as earlier from the Cape of Good Hope, he had returned to India convalesced but far from cured. During base-line operations in the Dun he again experienced great pain in his joints. A course of drastic treatment included the 'phlogistic diet', which may have killed Lambton, and much bleeding with leeches and 'cups' (the cup, or glass, was pressed against an area of scraped skin and heated so as to suck out 'bad' blood).

Somewhat recovered, in March 1835 Everest completed the connection of his Dun base-line to neighbouring stations. Then he again took to his sickbed. Four successive attacks of fever kept him there for the next six months, 'during which time I was once bled to fainting, had upwards of 1000 leeches, 30–40 cupping glasses, 3 or 4 blisters . . . besides daily doses of nauseous medicine, all of which produced such a degree of debility as to make it of small apparent moment whether I lived or died'.

His fate was, though, of increasing moment to his employers. There was now serious concern about whether he would ever be able to finish the Arc. 'I have survived the storm,' he announced in October 1835. But his recovery did not stop him reminding all and sundry that he was living on borrowed time, and lest he succumb again, he insisted on having Waugh by his side for the final triangulation. Meanwhile the government was sufficiently alarmed to start casting about for a possible successor.

News of this development, which Everest chose to interpret as an attempt to supersede him, would prove a greater restorative than any number of leeches. To scupper the appointment of Thomas Jervis, a man whom he deemed hostile to his achievements and scientifically unworthy of his post, Everest would pen and then publish his most vitriolic series of letters. If Jervis still fancied his chances, Everest would also demonstrate that rumours of his retirement were premature. To spite the wretched Jervis, he gritted his teeth and decided to soldier on, regardless of the consequences, until the Arc was finished.

By late 1835 the towers were ready. At the head of another large cavalcade Everest crossed the Siwaliks to begin the final triangulation between the Dun and Sironj. Needless to say, he found that the towers did not always conform to specifications. Those constructed by engineers from Delhi had been built of 'inadhesive materials', while instructions to isolate the pillar (for the instrument) from the gallery (for the observer) had been 'entirely lost sight of'. Agra's engineers had done much better, one of their towers being 'a perfect model of symmetry and elegance'.

For the most part, all that tedious work with flares and masts had paid off. Taking his angles in the early hours of the morning when refraction was at its greatest, Everest had little difficulty in sighting from one station to the next. The only setback came at the tower of Dateri, just east of Delhi, whence the sight-line to the next station on a ruined mosque at Buland-shahar proved to be blocked despite extensive tree-felling. It seemed, in fact, to be twice interrupted, once by a village called Ramnagar and then again by 'the lofty houses of the large town of Bhataona'. This was precisely the problem which all the preliminary triangulation had been designed to eliminate. There was now nothing for it but 'to cut a gap 30 feet wide' straight through both the village and downtown Bhataona.

To Rossenrode fell the unenviable task of placating the townsfolk and assessing the compensation to which the evicted would be entitled. 'How Mr Rossenrode has contrived to effect this severe operation . . . surprizes me,' wrote Everest. The people were Jats, agriculturalists with a reputation for extreme belligerence whom even well-wishers invariably described as 'sturdy'. It was, moreover, mid-winter. Being put out of one's house meant braving a heavy frost and bone-chilling fogs. Yet Rossenrode, venturing forth on his task unarmed and unsupported, somehow carried the day. In Ramnagar '5 huts, thatched, were crushed by the fall of the trees,' while in Bhataona '37 flat-roofed houses and 52 huts of mud [were] razed to the ground.' The hardship, as also the expense, was, in Everest's word, 'disastrous'. 'I hope,' he added, 'it will never again fall to my lot to have so disagreeable a task to discharge.'

South of Agra it was as if the populace had taken their revenge. Operating here not from towers but from the hill sites selected four years earlier, Everest found that many of his markers had been deliberately removed. It was further evidence of what he always called 'the suspicious native mind'. Like Lambton, he had already fallen foul of protective princelings anxious about the privacy of their womenfolk. In fact he could quite understand their concerns. An instrument which could turn women upside down ('an indecent posture, no doubt, and very shocking to contemplate') might also be able to see through things. It was, therefore, 'natural enough that they should assign to us the propensity of sitting all day long, spying through stone walls at those they deem so enchanting'. Likewise he was rather touched by the reverence sometimes extended to his instruments. The Great Theodolite attracted particular attention and, in backward areas like the rugged ravine country which he now encountered along the Chambal river, the instrument was much fêted by childless brides and other credulous supplicants. But he had no sympathy at all with those who, for much more understandable reasons, removed his markers.

The trouble seems to have stemmed from the mutual incomprehension which had come to characterise British–Indian relations. According to Colonel William Sleeman, a contemporary of Everest who was then famously engaged in suppressing the criminal caste known as 'Thugs', the Survey was causing deep rural anxieties. In particular its nocturnal habits and its predilection for hilltops, which were often the abode of a local deity, were deemed highly suspicious.

Needless to say, the choleric Everest was not the man to allay such superstitions, and nor, according to Sleeman, were the local Brahmins: 'The priests encouraged the peasantry to believe that men who required to do their work by the aid of fires in the dead of night on high places . . . must be holding communion with supernatural beings which might be displeasing to the Deity.' What more natural, then, than that pious locals should quickly exorcise the affront by digging up the embedded stone left by these unwanted sorcerers, or at least erasing the mystical mark which they had gouged in its surface.

Such wilful sabotage entailed the Survey in additional observations to relocate the original site, and then more laborious sinking of markers. The delays meant that it was impossible to complete the primary triangulation during the 1835–6 season. When the work was resumed in 1836–7, Everest compounded the delay by picking a quarrel with the authorities of the important state of Gwalior, through which the Arc passed between Agra and Sironj. Princely states in India, although not administered by the British, were invariably lumbered with a British Resident who acted as advisor and liaison officer to the state government, or *durbar*. The Resident in Gwalior, an exalted being in the coveted Political Department of British India, had a high regard for Gwalior's Maharajah, and had experienced some difficulty in convincing him that the Great Trigonometrical Survey should be made welcome in his territories. Everest thought he knew why. The Resident was incompetent; witness one of his letters to the Maharajah which referred to himself as 'one Major Everest engaged in measuring'. It was as bad as being called a 'compass-wallah'. Indeed it was 'the first instance of rudeness and opposition which I have experienced on the part of a British functionary'.

The Resident had since explained that, writing in Persian (the diplomatic language of India), he had been unable to find terms which could do justice to a title like 'Superintendent of the Great Trigonometrical Survey and Surveyor-General'. Everest remained incensed. He dashed off an official letter of remonstrance to Calcutta, he made the cardinal mistake of short-circuiting the Resident by appealing direct to the *durbar*, and he then found further cause for complaint when the promised escort of Gwalior troopers did not meet him on the state's frontier. For two weeks of the precious surveying season, the entire Great Trigonometrical Survey, about one thousand strong, languished on the Gwalior frontier while Everest waged his pointless vendetta. Since he was already provided with his own escort, the Gwalior troops were no more essential than had been the Hyderabad troops with whom he had come to blows nineteen years earlier in his first season with the Survey. But this time the outcome was different. When the Gwalior escort did eventually materialise, it was not they who were chastised but Everest. In no uncertain terms the government reprimanded him for assuming diplomatic status, wasting valuable time (its as well as his), and insulting one of its senior dignitaries.

Under something of a cloud, therefore, in 1837 the primary triangulation was at last carried south to Sironj. It was not a place which Everest recalled with affection. Back in 1824–5, the supposed insubordination of Olliver, the 'uncouth' language of Rossenrode, and the volubility of Rossenrode's horse had here driven him to the brink of insanity. Nor did he now have reason to revise his opinion of the place; for what should have been a major triumph with the completion of the Great Arc from one end of India to the other was marred by a fatal revelation. The base-line at Sironj as calculated by the triangulation carried down from the base in the Dun was found to differ markedly from that obtained by actual measurement over the ground in 1824–5. A couple of inches would not have mattered, but it was a case of a wholly unacceptable three feet. Something had gone badly wrong.

Everest's suspicions immediately focused on Dinwiddie's chain, the one used throughout by Lambton and the one on which he himself had relied for the Sironj measurement. If the error lay in the Sironj base, it could be discovered by remeasuring the base with the new compensation bars. But the bars were in store at Hathipaon. Trundling them the 450 miles down to Sironj would have to wait until after the 1837 monsoon.

Meanwhile an attempt was made to set up the two 'Astronomical Circles'. These instruments, specially made while Everest was in England, dwarfed even the Great Theodolite and were to be used for the final astronomical observations to establish the latitude and longitude of the Arc's extremities. But the test-run proved to be another dismal failure. When erected, the Circles were found to be insufficiently stable and had to be carted back to the Hathipaon workshops for modification.

By December 1837 the Survey and the compensation bars had made the forty-day trek back down to Sironj. Remeasurement of the base-line began immediately, but under Waugh's direction rather than Everest's. The Major's worst fears were being realised. Another catalogue of grisly symptoms had again confined him to his tent and its adjacent 'Necessary'. 'Dreadful rheumatic pains in my bones – fever – loss of appetite – indigestion – intestines totally deranged – stomach totally powerless – my strength entirely gone – the whole system apparently destroyed and for ever undermined.' He languished in his tent, not so much a caged lion as a cowed one. In marked contrast to the earlier measurement at Sironj, he declared himself hopeful, indeed touchingly confident, of his subordinates' abilities to conduct the operation on their own. In fact they were 'so thoroughly masters, each of his own part, that the measurement . . . proceeded just as satisfactorily as if I had been personally superintending it'. This was not the Everest of old. Weathered by age, achievement and ill-health, the Major was mellowing.

The compensation bars in due course produced a measurement for the seven and a quarter miles which differed from that obtained with Dinwiddie's chain by 2.79 feet. Thus the entire discrepancy, save for 6.395 inches, was accounted for. Everest would join in the general delight: 'considering that the Sironj and Dehra Dun bases are separated by nearly 450 miles and 86 principal triangles, [it] is as gratifying a proof of the accuracy of the series as could be desired.'

But to what extent he appreciated this triumph at the time is uncertain. Physically he was indeed recovering, but mentally Sironj was again taking its toll. To his 'indescribable dismay' he now found that not only was his eyesight affected but that his memory 'was in a great measure gone'. He was oppressed by 'a dreadful foreboding of ill'. It haunted his sleep and during waking hours took the form of 'some spectre or monster of the fancy coming to hold converse with me'. 'I thought it would certainly have ended in madness. Indeed I have little doubt that it would have . . . if I had not come to a better climate and foresworn business to a great extent.'

The better climate was that of Hathipaon and the Dun. Everest's days in the field were almost over. While he concentrated increasingly on supervising the astronomical operations and computing the results, the final observations for the Great Arc would be made by his sorely tried but now genuinely trusted assistants under Waugh as Assistant Superintendent.

The remaining tasks included the vertical triangulation of the entire series from Sironj to the Dun to establish the heights of all the stations and so of the Dun base-line, an important first step for the triangulation of the Himalayan peaks. Simultaneously re-observation of the triangles south from Sironj to

Hyderabad was undertaken and plans were laid to remeasure the Bidar base-line. To this end the compensation bars had been left in store at Sironj, suitably greased against corrosion with pig-fat and goose-lard and under the watchful eye of 2nd Assistant Owen Mulheran.

That Everest had not exaggerated his mental state may be inferred from the fate of the unfortunate Mulheran. After a few solitary weeks in Sironj, he was reportedly overcome by a fit of religious mania 'under which he successively burned off all his toes and several of his fingers in the slow fire of a candle'. Other manifestations of derangement 'of a similar and even more lamentable nature' included some wanton scratching of the precious bars. Fortunately they were not irrevocably damaged, and in 1841 were indeed employed to remeasure the Bidar base and thus complete remeasurement of the Arc to the same exacting standard all the way to Hyderabad, about nine hundred miles from the Dun. As for Mulheran, he too was not irrevocably discredited, although never entirely trusted. Four years later a colleague would draw attention to his curious habit of 'coming to office immediately after the internal and external application of a quantity of brandy and salt'.

The remeasurement of the Arc south to Bidar in Hyderabad had been undertaken to correct errors which might have resulted from the use of Dinwiddie's now disgraced chain and other inferior instruments. If Everest had had his way, the entire Arc would have been revised right down to Cape Comorin. But the government had only reluctantly approved the remeasurement of the Bidar base and could see no reason for further revision in the name of inch-perfect geodesy. Lambton's work was still good enough for all practical purposes.

There was even stronger resistance to Everest's unexpected suggestion of extending the Arc northwards. Following observations on The Chur and other nearby peaks, he had formed an ambitious plan of 'turning the flank of the mountains' by carrying the Arc up into western Tibet and on into Russian territory. The Chinese, who pretended to sovereignty over Tibet, would have to be persuaded to co-operate; but Everest thought that if both the Russians and the British could 'act combinedly', Peking's jealousy might be 'counteracted'. 'An arc of the meridian extending from Cape Comorin to the northern extremity of the Russian dominions near Nova Zembla!' he gasped. 'It is a vast project certainly! Utilitarians will scoff at the bare idea and say *cui bono?* ['to what good?'] Let these gentlemen prove to me the use of any earthly thing, and then I will take in hand to demonstrate the point at issue.'

The point at issue was, of course, the shape of the world. To Everest as to Lambton, discovering the precise figure of the earth was the most basic challenge in human science. It was a far greater 'desideratum' than, say, locating the source of the Nile or understanding the properties of electricity. Or, indeed, discovering the world's highest mountain. The Arc, quite apart from its cartographical, navigational and geological implications, promised the most intimate knowledge of the earth's dimensions; and if knowledge was the prerequisite of mastery, on it rested the future progress of man's management of his planet.

With the two great Astronomical Circles reinforced and installed in specially built observatories, Waugh at Sironj and Everest at Kaliana (near the northern end of the Arc but sufficiently removed from the mountains to eliminate their 'attraction') laboured simultaneously on forty-eight consecutive nights in December and January 1839–40 to observe some thirty-six pre-selected stars every night. In 1840–1 the same procedure was followed at Bidar and Sironj. So satisfied was Everest with these two sections of the Arc that he was pleased to note that there were now 'no two elements in nature more definitively known'.

Simultaneous observations of the same stars using identical

instruments and procedures was the surest way of getting precise comparative latitudes. From the grand total of over three thousand stellar observations, the latitudes of Bidar, Sironj and Kaliana were calculated to three decimal points of a second of a minute of a degree. The length of the Arc could now be deduced to the same standard of accuracy, and this value then correlated with the distance as computed by triangulation to obtain the 'amplitude' of the Arc.

In various reports and submissions Everest devoted reams of handwritten sheets to explaining his methods, to dilating in minute detail on the problems of refraction, plummet attraction and astronomical observation, and to recording his findings. A new set of constants – 'Everest's 2nd Constants' – were issued and showed that a semi-diameter of the equator at nearly twenty-one million feet exceeded the northern hemisphere's diameter by exactly 67,260 feet. The compression of the poles in terms of the diameter of the equator was thus 1:311.044.

But *cui bono?* indeed. Basically it was all numbers, page after page of angle tables and thirty-line equations involving every logarithmic device and geometrical formula known to mathematics. 'He undervalues everything that is not abstruse,' complained Sir Henry Lawrence, then a rising star in the administrative firmament. Instead of surveying India, the Surveyor-General sought only 'to astonish the savants of Europe'. The government wanted maps, or at least the co-ordinates for all Everest's trig stations on which they could be based. To them, as to most other people, all the rest was just too esoteric and too incomprehensible.

It was also too impermanent. Revisions seemed to go on indefinitely. At the mention of a new value for, say, refraction, or a new calculation of the co-ordinates for Madras, the whole thing required re-adjustment. The advent of the electric telegraph in the 1860s, and the opportunity this would provide for synchronising observations and so obtaining much more accurate readings for longitude, would constitute a veritable revolution in cartography and again necessitate extensive revision.

When in 1843, with the Arc completed, Everest finally put Hathipaon on the market and, embracing retirement, headed home, the Great Trigonometrical Survey was going from strength to strength. More regions, notably in what is now Pakistan, came under British rule and necessitated more chains of triangles. But of the Great Arc and its champion few traces would remain. 'No scientific man ever had a greater monument to his memory than the Great Meridional Arc of India,' wrote Sir Clements Markham, President of London's Royal Geographical Society; it was 'one of the most stupendous works in the whole history of science'. Yet for a total expenditure of about £150,000 the Great Arc had left precious little to show for itself. Sixteen weather-streaked towers still dotted the Doab, three largely deserted observatories in out-of-the-way places remained to puzzle the passing traveller, and atop a variety of *droogs*, hills and mounds several hundred station markers slowly succumbed to the combined assault of climate, vegetation and local prejudice.

Lambton's uninviting reports survive only in the dusty pages of *Asiatick Researches*, while Everest's two published accounts, though heavy and handsome, were poorly distributed and soon superseded. They are now unobtainable in all but a few specialist libraries. Much the most eloquent testimony to his life's obsession lies in the ruined shell of Hathipaon on its ridge above the Dun. There he spent his last years in India, dreading the health risks of a return to the plains, working on his reports and tables, and overseeing the operations of his subordinates. From the likes of Joshua de Penning in Calcutta, of Joseph Olliver and William Rossenrode, he now enjoyed the regard and affection which he had so often forfeited.

Both Rossenrode and Olliver had retired as soon as the Arc was finished but, with their sons and sons-in-law established in the Survey, they remained in close touch. As for old Joshua de Penning, the one-time incompetent and traitor, he had become 'my dear old friend'. De Penning would outlast even Everest, not retiring until 1845. In a letter from Calcutta of 1841 he fusses over Everest's health much like a loyal family retainer, and sends 'merino vests and drawers, a dozen of each ... packed up in four tin cases, which I hope will reach you in time for the cold season'. If Hathipaon has a ghost, he may be sporting woolly underwear. Perhaps, chanting logarithms from a windowless socket in what was once the drawing room, he gazes on the roofs of the Survey's Dehra Dun offices and then, swivelling like a theodolite, fixes unerringly on the spot known as Arcadia. With Hathipaon at the apex, the site of his terminal base-line on one side and the Dehra Dun head-quarters of the Survey on the other make as neat and evocative a triangle as any in India.

ELEVEN

A Stupendous Snowy Mass

In the mid-1830s, while Everest and Waugh had been putting the finishing touches to the Great Arc, four other parties from the Great Trigonometrical Survey had begun work on the 'bars' of Everest's cage-like 'grid-iron' of triangulation. In the 1840s, with the Arc complete, all resources were switched to this grid-iron and elaborate plans laid for its extension throughout the subcontinent.

Lambton's 'cobweb' of triangles in the south, though less neat and systematic than a grid, had provided the desired scatter of precisely located trig points from which cheaper topographical surveys could plot the detail needed for maps. The grid-iron was designed to furnish the same control for the rest of India, but with the trig points being arranged in 'bars' of triangulation.

A less contentious analogy was sometimes drawn from nature. Envisaging the north–south Great Arc as a tree-trunk, and its east–west limbs (like the Bombay and Calcutta longitudinal series) as branches, a tracery of slender fronds festooned with triangulated foliage (the 'bars' of the 'grid-iron') were to be superimposed on the subcontinent. Extending outwards from the Arc, the shade of their branches would define what the British deemed to be India in terms more organic (and so capable of further growth) and more congenial to tender consciences.

The immediate priority was to extend the control afforded by trigonometrical surveying to that part of northern India, the heartland of British rule, between the Great Arc in the west and Calcutta in the east. Joseph Olliver's seven-hundred-mile longitudinal series from Sironj to Calcutta, the one conducted during Everest's absence in England, provided the branch. Striking off from it at right angles, each 'bar' or 'twig' was to run north, roughly parallel to the Great Arc and at intervals of one degree longitude apart. With the Great Arc on the 78-degree meridian and Calcutta on the 89th, that meant eleven meridional series. Those at either end could be extended up into the Himalayas in the west, in the now British territories of Garhwal and Kumaon where Hodgson and Herbert had operated, and, in the east, into the Kingdom of Sikkim whose ruler permitted limited access to the area around Darjeeling. But for the most part the 'bars' terminated on the Nepal frontier, whence all approaches to the central Himalayas were still refused by the Kathmandu government.

With the exception of these extremities, the terrain over which the new tracery extended was that of the flat, densely populated and, in those days, generously shaded Gangetic plain. Here, because of the difficulties encountered by Everest with his flares and scaffolds, smaller triangles with shorter sides than those of the Great Arc were acceptable; but sight-lines had still to be laboriously ascertained and cleared, and innumerable stumpy towers erected. Nor was the work any less perilous. On the Nepal frontier, in the dreaded *terai* where Robert Colebrooke had once been stricken with malaria, whole survey parties now shared his fate. The death toll amongst both British and Indians sometimes reached three figures in a single season. The danger, wrote Clements Markham of the Royal Geographical Society, was 'greater than that encountered on a battle-field [and] the per-centage of deaths larger; while the sort of courage . . . required was of a far higher order'.

Casualties in the *terai* attended not just the eleven south–north series, which terminated amongst its wooded swamps and grasslands, but also a west–east series which, connecting the heads of the eleven meridional 'bars', was carried right through the *terai*. Known as the North-East Longitudinal series, it corresponded to Olliver's Sironj-Calcutta series at the southern end of the 'bars'. It would link their northern extremities by way of a 750-mile chain of triangles which ran parallel to the Himalayas from Everest's base-line in the Dun all the way to Assam.

Uniquely, the North-East Longitudinal was not, however, the work of one man or one party. None could have survived so many consecutive seasons exposed to its lethal conditions. Instead, each of the grid-iron's survey parties, having carried their triangles north to form one of the 'bars', then turned left to connect it up with the top of the next 'bar' one degree to the west. The North-East Longitudinal was thus pieced together over many years as each of the 'bars' was completed. It formed, as it were, the topmost branch of the whole tree. And in its carefully triangulated trig stations running along the base of the Himalayas there lay the long-awaited certainties from which the heights of the snowy peaks might at last be confidently observed.

Piling up the metaphors, Everest and then Waugh, his successor as Superintendent of the Great Trigonometrical Survey and Surveyor-General, conceived this Gangetic grid, or segment of the tree, as a quadrilateral. The 'bars' were contained within four sides consisting of the Great Arc itself, its two longitudinal branches (the Sironj-Calcutta series and the North-East Longitudinal) and an upright series linking them in the east known as the Calcutta Meridional. At each corner of this quadrilateral, accuracy was ascertained by a base-line measurement with the compensation bars. The bases at Calcutta, Sironj and the Dun had, of course, already been conducted by Everest himself. To complete the quadrilateral it remained only for Waugh to measure a fourth base-line in the far north-east.

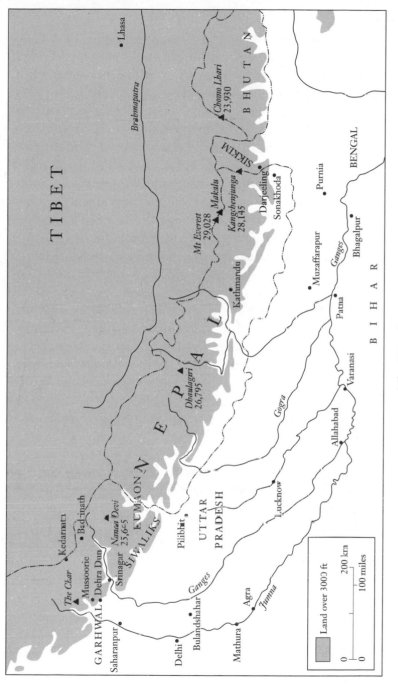

The Himalayas

The site chosen was at a place called Sonakhoda, below the Darjeeling hills where the North-East Longitudinal intersected the meridional upright carried up from Calcutta. There, in the moist plains of northern Bengal, Waugh and his assistants assembled with the compensation bars in late 1847. As with the Dun base-line, connection to the primary series, in this case the North-East Longitudinal, was made via stations on the neighbouring hills. It was while choosing and linking these, in the latter half of 1847, that Waugh found a new contender for the title of the world's highest mountain and so reopened the debate about the height of the Himalayas.

Everest himself had taken little interest in the subject. From the back door of Hathipaon he had been confronted by as fine a panorama of glistening summits as any in the world. They were good for the soul, but to his life's work on the Great Arc they were peripheral. From The Chur he may have actually sighted Nanda Devi; but there is no record of his having attempted to verify its height. Bagging mountain peaks was not his business. For those who had pursued the subject, often with inferior instruments and speculative observations, he felt only contempt.

Waugh, too, was circumspect on the subject. Although it was obvious that from the North-East Longitudinal series the secrets of the high peaks were within range, there was to be no unseemly rush to plot them. It was the sort of thing to which a surveyor might usefully devote his spare time while, say, waiting for towers to be built or trees cleared. Nor, when the peaks were indeed plotted, would there be any urgency to make public the results. The Survey had its code about publication, and no findings could be announced before exhaustive computation and revision of the data on which they rested.

The peak to which, almost casually, Waugh directed his theodolite while plotting the connection of the Sonakhoda base-line was Kangchenjunga, now perhaps the most easily observed of all the Himalayan giants and the third highest in the world. At the time Nanda Devi at the other end of the main Himalayan chain, the 'A2' of Hodgson and Herbert, was still credited with the greatest elevation yet measured. Webb's Dhaulagiri also had its champions, although the 28,000 feet once suggested for it by Henry Colebrooke had long since been dismissed as wildly improbable; something rather less than Webb's own estimate of 26,862 feet was thought more likely. In fact, it looked as if five vertical miles (26,400 feet) might constitute a pre-ordained ceiling above which no part of the earth was meant to protrude.

Waugh and Kangchenjunga now proved this wrong. But anyone who has seen Kangchenjunga loom from the clammy cloud-cover which envelops most of the eastern Himalayas for most of the year will find Waugh's encounter deeply unsatisfactory. A skilled and devoted professional, he lacked Everest's charisma and seemed content to live in his guru's shadow. Some found him sanctimonious; but if he did not endear himself to his subordinates, neither did he aggressively antagonise them. Fair to his mountains as to his men, Waugh eschewed comment as resolutely as Everest embraced it. Instead of penning narratives, he filed reports.

From above the hill resort of Darjeeling the dawn observer who is lucky enough, like Waugh, to beat the mist as it wells up from the Rangit valley enjoys one of nature's greatest spectacles. Forty miles away, across a chasm lined with rhododendrons and bubbling with cloud, the mountain stands detached from the ground and seems not of this world. Rather does it materialise, ghost-like, out of the lightening sky. You look for it on the horizon and find that you have been staring into its navel. The summit, cleft by a wall of granite and defined by its glistening flanks, sails high overhead like a celestial Olympus etched in chill sunlight.

'The western peak of Kangchenjunga attains an elevation of no less than 28,176 feet above the sea, which far exceeds what has hitherto been conjectured,' wrote Waugh in ink as dry as dust. He and his assistants, including William Rossenrode junior, had observed it from Tiger Hill, Senchal, Tonglu and most of Darjeeling's other now renowned viewpoints. It was much the highest known mountain in the world, being nearly three thousand feet in excess of Nanda Devi. And since it had been approached more closely than any of its rivals, and from a base-line subject to the rigorous controls of the Survey, the observations could be taken to be unassailably accurate. Though incidental and unexpected, Kangchenjunga's primacy could be seen as a crowning triumph for the Great Arc.

Yet Waugh did not announce this discovery until two years later. Even then he did so only in an internal memorandum; for doubts had arisen, not about Kangchenjunga, but about another peak to which, from Darjeeling, he had also taken bearings. The bearings did not include vertical angles because the peak in question was deemed too distant and indistinct. Like other such irregularities on the horizon, its position was plotted, its profile sketched, and it was then given a sequential designation. Waugh used the letters of the Greek alphabet. The distant peak, lying to the left of Kangchenjunga and at least 120 miles away on the Nepal–Tibet border, became 'gamma'; and although loath to admit it, he already suspected that 'gamma' might exceed Kangchenjunga.

Waugh conducted his Darjeeling observations in November 1847. In the same month, but from the North-East Longitudinal at Muzaffarapur in Bihar, John Armstrong, one of the many assistants recruited by Everest, had taken three sets of horizontal angles and one vertical angle to a shy and partly obscured giant which, as Waugh immediately suspected, proved to be the same mountain as 'gamma'. Armstrong had listed his peak simply as 'b'; and from his angles, a height of 28,799 feet seemed to be indicated. But 'on account of the great distance', Waugh distrusted Armstrong's observations as much as his own. He decided to await the outcome of the

1848–9 season. 'I particularly wish you to verify Mr Armstrong's peak "b",' he told John Peyton, once one of Everest's prized 'computers'. 'His [Armstrong's] peak "a",' Waugh added, 'also requires to be well verified because the two heights deduced are very discordant.' Almost certainly, this 'a' was Makalu, today reckoned to be 27,805 feet and so the fourth highest in the world. It stands on the Nepal–Tibet frontier just to the east of a cluster of giants including the timid fang which was Armstrong's 'b' and Waugh's 'gamma'.

Peyton had no joy in 1848–9. The peaks were visible only in the early mornings and only during November and December. Of a morning, by the time his instrument had been trained on them, they had disappeared; and of a season, by the time his survey towers had been built, the peaks had gone into hibernation behind a veil of cloud which lifted not even at daybreak. Primarily concerned with contributing his section to the North-East Longitudinal, Peyton found it impossible to have towers ready early enough in the season for mountain triangulation.

A year later, with more encouragement from Waugh, James Nicholson succeeded Peyton and resumed the quest. Edging east, the North-East Longitudinal took Nicholson slightly closer to the target. His 'sharp peak "h" ' was clearly Armstrong's 'b' and Waugh's 'gamma', and he concentrated his attentions on it. Numerous angles, both vertical and horizontal, were taken, six of which were used in the final computations. But although Nicholson himself must have known the outcome by early in 1850, Waugh was in no hurry to proclaim it.

All the Himalayan peaks were first given new designations, this time in Roman numerals from I to LXXX. 'Gamma'/'b'/'h' now became Peak XV. Waugh then, in the words of Reginald Phillimore, the Survey's historian, 'asked the Chief Computer in Calcutta to revise the form [formulae?] for computing geographical positions of snow peaks at distances of over 100 miles'. The Chief Computer was Radhanath Sickdhar, the

Bengali genius whose arithmetical wizardry had so impressed Everest. A later tradition, dismissed by Phillimore although accepted by many Indian historians, that it was in fact Sickdhar who first realised that XV was the world's highest presumably stems from this reference. The popular account of the excited Bengali rushing into Waugh's office exclaiming that he had 'discovered the world's highest mountain' is obviously rubbish. Waugh's office was in Dehra Dun while Sickdhar was now in Calcutta. But it is quite probable that Sickdhar's computations provided the first clear proof of XV's superiority.

'For the next four years,' continues Phillimore, '[Waugh] was discussing refraction coefficients and the datum zero height which had to await tidal observations at Karachi.' Then 'as a final check he wrote for the old records of Charles Crawford and William Webb.' It was not, therefore, until March 1856 that Waugh at last took up his pen and, in a letter consisting of fourteen numbered and neatly written paragraphs, summarised his findings.

The letter, 'No 29B', might be 'made use of' but it was not for publication. The results were still provisional; there was much revision yet to be undertaken. It was addressed simply to Captain Thuillier, his Deputy Surveyor-General in Calcutta. But its contents were such that they quickly became common knowledge. For after a four-paragraph preamble, Waugh at last directed his attention to Peak XV.

> 5. We have for some years known that this mountain is higher than any hitherto measured in India and most probably it is the highest in the whole world.

> 6. I was taught by my respected chief and predecessor Colonel Sir Geo. Everest to assign to every geographical object its true local or native appellation. I have always scrupulously adhered to this rule as I have in fact to all other principles laid down by that eminent geodesist.

7. But here is a mountain, most probably the highest in the world, without any local name that we can discover, whose native appellation, if it has any, will not very likely be ascertained before we are allowed to penetrate into Nepal and to approach close to this stupendous snowy mass.

8. In the meantime the privilege as well as the duty devolves on me to assign to this lofty pinnacle of our globe a name whereby it may be known among geographers and become a household word among civilized nations.

9. In virtue of this privilege, in testimony of my affectionate respect for a revered chief, in conformity with what I believe to be the wish of all the members of the scientific department over which I have the honour to preside, and to perpetuate the memory of that illustrious master of accurate geographical research, I have determined to name this noble peak of the Himalayas Mont [*sic*] Everest.

10. The final values of the co-ordinates of geographical position for this mountain are as follows, viz –

MONT EVEREST OR HIMALAYA PEAK XV

Latitude N	Longitude E of Greenwich	Height above sea level
27° 59' 16.7''	86° 58' 5,9''	29002 feet

As intended, Thuillier duly conveyed this information to members of the Asiatic Society in Calcutta. They approved Waugh's findings although not the new name. The latter, which Waugh himself quickly changed from 'Mont Everest' to 'Mount Everest', was however endorsed in London by the Secretary of State for India and by the Royal Geographical Society.

Doubts, though, remained. For one thing, it was not certain

that the new peak was in fact the highest. By the time Waugh composed his letter, British India had devoured the lands which today comprise Pakistan. Leafy branches of triangulation were spreading rapidly west and north-west, particularly into the newly created state of Kashmir, whose uncertain mountain borders marched with those of China and several central Asian kingdoms. The latter were rapidly succumbing to Russian influence, and with the British paranoid about Tsarist designs on their Indian empire, the mapping of Kashmir had been given the highest priority. In 1856, even as Thuillier was conveying Waugh's news of Mount Everest to the Asiatic Society, a party of shivering surveyors was encamped on Haramukh, the mountain which presides over the Kashmir valley. From there they were taking angles to a new cluster of peaks, distant 140 miles and evidently of exceptional magnitude.

The range in question, detached from the chain of the Great Himalaya and just to the north of its western bastion (the 26,660-foot Nanga Parbat), was said to be called the Karakoram. Captain Montgomerie, the man in charge of the Kashmir survey, therefore numbered its peaks with the prefix 'K' and, in a small sketch, clearly delineated the first two in the new series. For his 'K1' a local name was later found – Masherbrum. 'K2', a sharper and more elusive peak, remained anonymous although not ignored. The possibility that it might exceed Mount Everest was clear in 1856 and led to a succession of observations in 1857, then some hasty computation in 1858. Montgomerie, unlike Waugh, was keen to dispose of the matter, although probably disappointed by the result. At 28,287 feet (later revised to 28,168), K2 was slightly higher than Kangchenjunga but well short of Mount Everest.

Other challengers would also be seen off. The altitude of Mount Everest has since been often adjusted but seldom to below 29,000 feet. At either 29,028 or 29,141, it reigns supreme. But this supremacy only fuelled another debate: why should it be called 'Mount Everest'? 'K2', for instance, remains

'K2'. Names likes 'Keychu' and 'Keytu' would be exposed as no more than local renderings of Montgomerie's designation; other names, including 'Mount Waugh' (after the Surveyor-General), 'Mount Albert' (after Queen Victoria's consort), 'Mount Montgomerie' (after its 'discoverer') and 'Mount Godwin-Austen' (after the surveyor who first actually penetrated the Karakoram), have failed to win acceptance. No doubt the government of India or that of Pakistan would happily adopt a new name for the peak but, while control of Kashmir continues to be disputed, K2 is likely to remain a nameless orphan.

The case for scrapping 'Mount Everest' rested on the suspicion, anticipated by Waugh, that there might be a local name for it which access to Nepal would reveal. Brian Hodgson, an eminent Buddhist scholar who had resided at Kathmandu for some years, immediately came up with 'Devadhanga' as the Nepali designation. The Asiatic Society, deferring to Hodgson's scholarship and reflecting the hostility which many in British India still felt for George Everest, agreed. But Waugh objected. He convened a committee which declared Devadhanga 'indefinite and unacceptable'. Although enshrined in Nepali legend, it apparently applied to several peaks. In the past such imprecision had scarcely deterred adoption of a name; but the fact that XV was the world's highest, that Waugh had already named it, and that 'Mount Everest' was indeed rapidly becoming 'a household word among civilized nations' militated against change.

So did the turmoil which swept northern India in 1857. Within a year of Waugh's announcement, the British were fighting for the existence of their Raj. In the context of what they insisted was just an 'Indian Mutiny' but which Indians regard as a great national rebellion, the quibbling over the name of a mountain abruptly ceased.

The Great Rebellion, though sparked by a mutiny of Indian troops, spread across a national landscape parched by years of withering contempt for the sensibilities and customs of India's people. It would be unfair to claim that the Rebellion, like the measurement of Mount Everest, stemmed from the activities of the Great Trigonometrical Survey. But surveyors had undoubtedly fuelled both the British sense of superiority and the Indian sense of grievance. 'Bars' and 'chains' of invisible triangulation looked and sounded a lot like political strangulation. Not unwittingly the Survey had furnished the paradigm and encouraged the mind-set of an autocratic and unresponsive imperialism. Additionally, by razing whole villages, appropriating sacred hills, exhausting local supplies, antagonising protective husbands and facilitating the assessment of the dreaded land revenue, the surveyors had probably done as much to advertise the realities of British rule and so alienate grassroots opinion as had any branch of the administration. Back in the early 1800s, men like Mackenzie and Lambton had respected and even admired India's rich cultural traditions. But to Everest and his generation devotional customs and immemorial lore were just evidence of 'the suspicious native mind'. Tiptoeing round local sensibilities, whether Indian or British, was not an art which George Everest had ever recognised.

With the British in India otherwise engaged, 'Mount Everest' won international recognition. When the name was again questioned, the logic of sticking with it was stronger than ever. In the early twentieth century the great Swedish explorer Sven Hedin had come up with a long Tibetan name for the mountain. Rendered in various different spellings, 'Cha-mo-lung-ma' was also, like Devadhanga, rejected on the grounds that it was applicable to the whole Everest region rather than to a particular peak. Still longer Tibetan names like *Mi-thik Dgu-thik Bya-phur Long-nga* (which one writer translates as 'You cannot see the summit from near it, but you can see the summit from nine directions, and a bird which flies as high as the summit goes blind') are undoubtedly more specific. But they scarcely trip off the tongue, nor do they endear themselves to cartographers working within the cramped confines of a small-scale map. 'Mount Everest', on the other hand, universally mispronounced and long since disassociated from its contentious namesake, has a ring of permanence, an aura of assurance.

Strangely, the one person who might have entered into this debate with intriguing effect held his peace. It was not out of modesty. George Everest, after declining one order of knighthood because he thought it not grand enough, had in 1861 become Sir George Everest, Companion of the Bath. In a typically overblown disclaimer he had once confessed to being 'by no means disposed to be very humble, or to play the courtier, or to kiss the rod that chastises me'. Yet of his reaction to having the world's highest mountain named in his honour there is no record at all. Perhaps he rightly judged that any intervention on his part might be counter-productive.

For a man who had been far from well for the past twenty years, Everest's homecoming had had a dramatic effect. Reaching England in 1844, he settled first in the Leicestershire countryside, where he was soon riding with the local hunt, and then in London. In 1845 he visited the USA, and in the following year, back in London, he married. He was then fifty-five. To Everest, as to Lambton, the joys of family life constituted a last great discovery. Seemingly his bride shared this sense of achievement. Although Emma Wing was less than half his age, she proved to be a devoted wife who over the next ten years bore him six children.

The last glimpse of the great man, as later recollected by his eldest surviving son, reveals a contented old gentleman, friendly with the explorer David Livingstone, the chemist Michael Faraday and other notable contemporaries. Adopting that leonine beard and hairstyle, he enjoyed the plaudits of the scientific societies but was just as content playing the Victorian father. The day began with family prayers 'at which the servants attended'. 'My father was a firm believer in God as every

Freemason ought to be.' There might follow a few hours' work, perhaps on a mouth-watering paper like that 'On Instruments and Observations for Longitude for Travellers on Land' (published in 1859), and then a lecture at the Royal Institution.

Most days there was also time for a bit of parental instruction. With his offspring perched on high stools at a long deal table, he introduced them to the mysteries of elementary arithmetic, algebra, geometry, trigonometry, and 'learning something about logarithms'. Perhaps, too, there were tales about the tigers he had never met and the mountain he had never seen. He died in London in 1866, aged seventy-six, and was buried in Hove, near Brighton.

No statue has ever been erected in his memory. George Everest, like William Lambton and like their Great Arc, was soon forgotten. But where history is oblivious, geography is tenacious. By having, in the words of Waugh's successor as Surveyor-General, 'placed his name just a little nearer the stars than that of any other lover of the eternal glory of the mountains', the maps continue to acknowledge their debt to the ever-restless genius of George Everest.

A Note on Sources

Anyone familiar with R.H. Phillimore's *Historical Records of the Survey of India* (5 vols, Dehra Dun, 1950–68) will recognise my principal debt. Without Colonel Phillimore's monumental, if eye-straining, digest of the Survey's records, this book could scarcely have been written. Phillimore's volumes I–IV are available in many libraries but volume V, which deals with the period 1843–60, was withdrawn because of the strategic sensitivity of some of the subject-matter. Only three copies are known to exist in the UK – one each in the British Library and in the libraries of the Institute of Chartered Surveyors and of the Royal Geographical Society. Clements R. Markham, *A Memoir of the Indian Surveys* (London, 1871), has a useful map of the Great Trigonometrical Survey, but has otherwise been superseded by Phillimore.

The Asiatic Society of Bengal's *Asiatick Researches*, vols VI–XIV (Calcutta, 1804–22), contain Lambton's reports, mostly of a technical nature. Vol. XII includes Henry Colebrooke's paper 'On the Heights of the Himalaya Mountains', Vol. XIII has Webb's memoir on his Kumaon survey, and Vol. XIV the findings of Hodgson and Herbert in Garhwal. The reports of Crawford's observations in Nepal and the extracts from Robert Colebrooke's diary are as per Phillimore vols II–III. The Godfrey Thomas Vigne extract is from *Travels in Kashmir, Ladakh, Iskardo* etc. (London, 1842).

The *Quarterly Review*'s critique of Colebrooke's paper appears in its July 1817 issue in Vol. XVII; its retraction is tucked away in a review of Alexandre de Humboldt's *Sur l'Elevation des Montagnes de l'Inde* in the January 1820 issue in Vol.

XXII. For Playfair's review of Lambton's work see the *Edinburgh Review* of July 1813 in Vol. XXI. And for James Prinsep's account of the Calcutta base-line see *Journal of the Asiatic Society of Bengal*, Vol. 1 (Calcutta, 1832).

Most of the Everest extracts are from his *An Account of an Arc of the Meridian* (London, 1830), *An Account of a Measurement of Two Sections of the Meridional Arc* (2 vols, London, 1847) and *A Series of Letters Adressed to HRH the Duke of Sussex* (London, 1839). These contain Everest's own, not impartial accounts of his work. They have been supplemented by reference to Phillimore's extracts from his correspondence in the Survey's archives.

The bicentenary of Everest's birth in 1990 occasioned a couple of symposia which resulted in the Survey of India's *Souvenir of the Birth Centenary of Col. Sir George Everest* (Dehra Dun, 1990) and the Royal Institute of Chartered Surveyors' *Colonel Sir George Everest: A Celebration of the Bi-centenary of his Birth* (London, 1990). Papers on the life of Everest by J.R. Smith, on the triangulation of the Cape of Good Hope by Colin Martin and Roger Fisher, and on map-making policy in India by Matthew Edney were found particularly relevant.

On an earlier occasion, the first ascent of Mount Everest in 1953, a useful summary of the 'Heights and Names of Mount Everest and Other Peaks' by J. de Graaff-Hunter appeared in *Occasional Notes of the Royal Astronomical Society*, No. 15, October 1953.

Other works which proved helpful include: Simon Berthon and Andrew Robinson, *The Shape of the World* (London, 1991); Matthew Edney, *Mapping an Empire: The Geography of India* (London, 1997); J. Howard Gore, *Geodesy* (London, 1891); Arthur R. Hinks, *Maps and Survey* (Cambridge, 1913); Kenneth Mason, *Abode of Snow* (London, 1955); W.A. Seymour (ed.), *A History of the Ordnance Survey* (London, 1980); R. Smyth and H.L. Thuillier, *A Manual of Surveying for India* (Calcutta, 1851); John Noble Wilford, *The Mapmakers* (London, 1981).

Finally, some details have been drawn from three of my own books: *India Discovered* (London, 1981 and 1993), on Mackenzie and the early surveys; *The Honourable Company: A History of the English East India Company* (London and New York, 1991), on the political background; and *When Men and Mountains Meet* (London, 1977), reprinted in *The Explorers of the Western Himalayas* (London, 1996), on the Kashmir Survey.

Index